# Thyroid: The Butterfly of Metabolism

Daniela Pace

# Thyroid: The Butterfly of Metabolism

How to prevent, take care of oneself, and stay healthy

 Springer

Daniela Pace
Endocrinology Department
Hospital San Carlo di Nancy
ROMA, Italy

ISBN 978-3-031-55275-5          ISBN 978-3-031-55276-2   (eBook)
https://doi.org/10.1007/978-3-031-55276-2

English translation of the original Italian edition published by E.L.I. Medica, Villaricca, 2022

The translation was done with the help of an artificial intelligence machine translation tool. A subsequent human revision was done primarily in terms of content.

Translation from the Italian language edition: "Tiroide: la farfalla del metabolismo" by Daniela Pace and Elisabetta Colagrande, © E.L.I. Medica 2022. Published by E.L.I. Medica. All Rights Reserved.

Illustrations by Elisabetta Colagrande

This Springer imprint is published by the registered company Springer Nature Switzerland AG
The registered company address is: Gewerbestrasse 11, 6330 Cham, Switzerland

If disposing of this product, please recycle the paper.

*To those willing to understand and delve into medicine through texts based on evidence and the achievements of science.*

*Those who have a hunger for knowledge—meant as a tool to address doubts and concerns—are aware that it is the only winning weapon against loneliness, anxiety, and despair of illness.*

# Foreword 1

*"The really important thing is not to live, but to live well".*
Plato

Divulging medical-scientific information, promoting prevention, and providing precise answers about the functioning of our organs, the diseases that may affect them, and the correct lifestyle for their protection are some of the primary functions of every good physician. These functions are particularly crucial today, in a world that changes at the speed of light, where technological innovation and the increase in life expectations profoundly alter our relationship with our health and approach to care.

Until a few decades ago, the doctor–patient relationship was essential and much more direct. However, today, in the presence of suspicion or the onset of the first symptoms of a disease, the web and social networks have become the true reference point for every generation, from Gen-Z to seniors.

This has given rise to the so-called DIY (do-it-yourself) medicine, the unhealthy practice of self-medication, which only throws us into that sense of confusion very much representative in situations of discomfort. That is precisely why, now more than ever, an effort for constant communication is essential to strengthening the authority of the medical profession as well as bringing its role back to the centre.

Medicine, however, must be introduced in a more accessible way and available to everyone, increasingly assuming the same immediacy and characteristics of the interconnected and intermedial society in which we live. This is the only way the balance can be restored and allows each and every one of us to consciously handle our health, becoming an active subject who is motivated to act in respect of our well-being and encouraged to eliminate any incorrect behaviour from our lives; health lives in every place and moment of our day, taking care of ourselves as well as the others, and above all, making ourselves the main character of life's circumstances.

The idea of this book is born from these considerations, along with the desire to convey and acquire knowledge, meant as the primary tool for prevention or—in the case of diagnosed pathology—reaction and counteraction to the disease.

With *Thyroid: The Butterfly of Metabolism*, the author shares information about this gland, as small as it is crucial for our bodily functions, about the pathologies that concern it, diagnostic techniques and procedures, treatments, and the behaviour to adopt in order to preserve it from dysfunctions. She does so with an approach and a lexicon that makes the text accessible to a vast audience, avoiding the medical concepts distortion, but making scientific protocols, explanations, and therapeutic solutions more streamlined.

In this book, we meet Dr. Pace, her academic background, and her activity as an endocrinologist expert in neck ultrasound. We feel her experience in scientific research and teaching at the Advanced Training School in Thyroid, Parathyroid, and Neck Pathologies Ultrasound of the Italian Society for Ultrasound in Medicine and Biology (SIUMB).

At the same time, we meet Daniela, her personal journey, from her past as an oncology patient, growing and developing, to the moment she embraced the medical profession with a constant desire to improve herself in order to better serve others: a vocation she feels strongly and puts into practice through various volunteer activities. To her, the white coat represents the highest expression of altruism and embodies the highest values of human solidarity, the pure and simple one of those who smile at life by silently donating their time. This is why it does not come as a surprise that she is engaged as the organizer of *ThyroiDay*. This is an initiative that is dedicated to the prevention and awareness of thyroid disease diagnosis (the experience was interrupted in 2019 due to COVID-19). Nor should it be her active contribution to the Velletri Municipal Blood Donor Association during Thyroid Screening Days, a study program she conceived and curated as part of the Thyroid Disease Prevention Campaign, aimed at identifying potential issues in donors.

In the following pages, both the doctor and the former patient emerge in all their energy: the study years, research, and professional updating of Dr. Pace forcefully intersect with Daniela's personal and clinical history. Together, they reveal, alongside medical-scientific explanations, that unique empathy that only those who have experienced certain situations first hand can have.

This is how the words in this book become an informed caress for us readers, an important opportunity to take care of ourselves and our loved ones, well aware that health—quoting the Ottawa Charter—is "a resource for daily life" and not "the goal of living".

Enjoy your reading!

Fabrizio Giona
Journalist and Science Communicator
Italy

# Foreword 2

When involved in the world of medical and scientific divulgation, one learns that when one speaks, clarity is necessary. It does not matter whether you speak a lot or a little: what matters is that everything has to be precisely said, also making the listener or reader feel at ease.

The listener should never perceive a sense of confusion; we should not need to reread or re-listen the things after they are explained to us. On the contrary, we should be able to say with a certain relief, "Yes! I get it!" Only in this way we can acknowledge, understand, and therefore be consistent with therapies even if they may be a bit unpleasant. We should indeed trust the person who is speaking to us since they genuinely want us to understand the problem and, above all, want to help us overcome it—or even, hopefully, induce prevention.

In Daniela Pace's book, two elements stand out: the lightness of the writing and accuracy. It is confidential, deliberately discursive when it needs to be intriguing, and rigorously scientific when it wants to convey correct knowledge, free from any brazen simplification.

In short, a dual register ensures that the young person, or the novice, does not feel detachment or disinterest in what the author tells about the thyroid. It also ensures that the medical student properly reviews, or the patient who may have already had issues with this small yet "turning up like a bad penny" gland recognizes and delves deeper; even understanding better, in short, where they may not have been able or willing to directly ask questions.

The illustrations make a great contribution to strengthening the underlying project of this concise work. Explaining everything without weighing it down.

A virtuous path can be generated, beginning from the seriousness of science but then mostly focusing on its humanity. After all, medicine is not born to show off its erudition but to help people to live better and longer, thanks to the healing power of prevention and, on top of that, the power of medicines, therapies, and technology.

No physician heals their patients with words alone, although words chosen correctly may create the most agile path to achieve the goal of every good doctor: all their patients are in good health and stay healthy.

Daniela Pace has clearly understood this. With this effort, she not only demonstrates to be a good writer but beyond everything, a good doctor.

Benedetta Rinaldi
Journalist and Science Communicator
Italy

# Foreword 3

It is with great pleasure that I applaud the excellent work carried out by Dr. Daniela Pace, who deserves credit for contributing to the debate on the thyroid and the indispensable role of the physician in scientific promulgation and the promotion of well-being. Her contribution as a result of extensive field experience fully emerges from the pages of this book and suggests to all of us the need to lower the "tones" of medical communication, making it accessible and practicable for everyone.

The thoroughness with which the colleague delves into explanations is commendable, leaving nothing implicit and particularly avoiding—as it often happens among professionals—the self-referential need to demonstrate knowledge. In the following pages, there is a desire to inform and make people aware, always keeping in mind the three fundamental pillars of the medical profession: studying, analysis, and sharing.

This publication represents a well-structured itinerary that analyses thyroid issues and illustrates the causes, possible treatments, and procedures. The ultimate goal is to bring the importance of knowing the morphology and functionality of the thyroid back into the debate, aware of how very common diseases associated with it are in the population, a number that is constantly increasing, affecting individuals of both sexes at all ages, with a higher incidence in women. Among these pathologies, a significant issue in our country is represented by thyroid nodules, mainly a consequence of insufficient iodine intake. This deficiency leads to an enlargement of the thyroid, known as goitre. Finally, the malignant neoplasms of the thyroid should not be underestimated, representing 3–4% of all human tumours. They mainly affect women, especially between 40 and 60 years old, with an incidence of about 15–18 cases per 100,000 women (for men, there are about five cases per 100,000 male inhabitants).

As an Endocrinologist and President of the Italian Thyroid Cancer Observatory Foundation (ITCO), I cannot help but be grateful for this important, informative, and educational work. Reading these pages helps develop particular attention to the thyroid, singularly navigating its various functions and dysfunctions, stimulating the adoption of healthy and sustainable behaviours for our body. Also, it indirectly encourages the creation of a multidisciplinary network of specialists capable of better measuring and understanding reality and, therefore, guiding the patient more effectively in prevention and, where necessary, towards adequate treatments.

Cosimo Durante
Italian Thyroid Cancer Observatory
Foundation, Rome, Italy

# Foreword 4

Many tend to associate thyroid dysfunctions only when they see weight fluctuations, and of course, they are unaware that, despite its small size, the gland regulates numerous functions in our body. Therefore, its malfunctioning is observable through an extensive and diverse range of symptoms. Some may be immediately detectable through simple blood tests, while others may be subtler and can only be identified by specialists with the help of various diagnostic techniques, including ultrasound.

Using ultrasound allows the endocrinologist to obtain a precise image of the thyroid, providing important information about its structure and morphology. It also enables a detailed analysis of any thyroid nodules, and Colour Doppler provides data on the vascularization of the gland and any focal lesions. It is clear that the advent of ultrasound and the technological and medical-scientific development over time have undoubtedly and profoundly changed the diagnostic process of thyroid pathology, offering more precise and prompt possibilities for intervention, treatment, and prevention.

Dr. Daniela Pace, who is also a lecturer at the *Advanced Training School in Thyroid, Parathyroid, and Neck Pathologies Ultrasound* of SIUMB, is well aware of the importance of this diagnostic technique and the advantages it offers, including other methods such as fine-needle aspiration or procedures such as *alcoholization* and *thermoablation*. Her sensitivity, combined with professionalism and a particular dedication to studying and keeping updated, decisively contributes to the development of the debate and education in this field, emphasizing the fundamental role of ultrasound as a non-invasive diagnostic tool. It also underlines the importance of entrusting it to experienced operators with a solid clinical-specialist background.

Publications like the one we are presenting are important in this perspective as they highlight aspects that sometimes go unnoticed and, in particular, allow their extension to an audience who is not involved in the field, making them perfectly aware of the possibilities that technical and scientific developments offer in the matter of diagnosis and treatment of thyroid pathologies.

We cannot but welcome this book with great enthusiasm, thanking Dr. Pace for the excellent work done in terms of promulgation. A text that deserves to be read and consulted throughout our lives, as it reminds us of how important it is to take care of ourselves, continuously focus on our psychophysical well-being, and live in full awareness.

Giovanni Iannetti
Stefano Spiezia

# Foreword 5

Thyroid disorders are very common and can affect individuals of both sexes, with an impact on health and quality of life that can vary based on the degree of dysfunction, age, and the condition of the affected individual. The consequences can be more significant and disabling if they occur at specific times, such as during the development stages of pregnancy.

Endocrinology data outline clear scenarios and contexts, with the incidence of thyroid diseases constantly increasing, reigniting the debate on the role of the national healthcare system regarding this significant incidence of morbidity, lethality, and mortality.

Scientific research and constant technological development undoubtedly lead to new and progressively specific intervention possibilities, seeking to provide precise and conclusive answers to doctors, specialists, and patients. However, the discussion around the concept of "medical culture" makes the difference: this primarily concerns healthcare professionals and operators, not only in terms of knowledge and how to use it but also in terms of a marked ability to address the "medical question" from a divulging perspective. This is the keystone that even facilitates the redefinition of the domains of medicine and the clinical method. It is the art of oratory, so dear to the Roman philosopher Marcus Tullius Cicero, expressed in the sense of *docere*, meaning to clearly inform and demonstrate a thesis most convincingly.

This text is published to inform and divulge medical knowledge, specifically about the thyroid and everything one needs to know to prevent or treat the diseases it is subject to.

Dr. Daniela Pace, a colleague, follows a precise oratory path, managing to translate pathophysiology, clinical culture, and statistical-analytical culture into thorough content, albeit simplified in both linguistic and procedural terms. The primary objective is to promote the full participation of the patient in the informational process, so that they can become fully aware and capable of facing a preventive—primarily—or therapeutic path when faced with a confirmed diagnosis of pathology.

The goal is noble and undoubtedly deserves the praise of the scientific community for its strong social value. As an Ordinary Professor of Endocrinology and holder of the UNESCO Chair at the University of Naples *Federico II*, I cannot but appreciate, beyond the content, the spirit of sharing that has generated this publication and the expected outcomes, with particular reference to the educational mission aimed at younger generations, both in terms of prevention and the promotion of healthy lifestyles.

Annamaria Colao
University of Naples "Federico II", Naples, Italy

Health Education and Sustainable
Development, UNESCO, Paris, France

# Foreword 6

The doctor–patient relationship has undergone, especially in the last decade, a significant evolution, changing substantially the rigid roles played in the past by the interlocutors. As a matter of fact, the doctors used to position themselves as conservators of inaccessible knowledge, and the patients were unaware beneficiaries who entrusted themselves to some care in a less participatory manner, "undergoing" the choices made by the caregiver in their interest.

Regarding the perception of health and the approach to prevention and medical care, contemporary society has taken giant steps, especially thanks to the continuous communication evolution, which has increasingly become a crucial tool not only for information but also for the true care of the patient.

In a complex and technologically advanced context like contemporary medicine, the real challenge for the doctor lies in the ability to produce simplified but scientifically correct communication models that make medical culture accessible and shared. This enables the patient to become an active and involved participant, capable of making choices for their health with full autonomy and freedom of judgement.

This is the so-called empowerment referred to by the World Health Organization (WHO), and it forms the foundation for the following pages. Through them, Dr. Daniela Pace virtually takes us into her office, illustrating step by step this incredible gland called thyroid, which regulates many functions of our body.

Guiding this publication is her undisputed experience as an endocrinologist, along with empathy, and, last but not least, an awareness of the strategic role of communication in medicine, exactly as highlighted by the Code of Medical Ethics, which establishes it as the foundation of the doctor–patient relationship, considering it "time for care".

I personally had the honour and pleasure of seeing Daniela in action, both in her tireless daily clinical activity and in the passionate writing of this book. Therefore, I can rightly confirm the commitment she invests in communication, based on a deep and constantly updated knowledge of the subject, as well as her innate ability to make it "understandable and comprehensive", just as the Code of Medical Ethics envisions.

The pages we are about to read are precisely so: the graphic and descriptive representation, understandable and enjoyable, of the exceptional work that Daniela does every day in her office, guiding patients to take care of themselves, teaching them how to recognize the signals sent by their own bodies, and encouraging them to be aware of the importance of prevention and participation in their care.

Emanuela Traini
Endocrine Surgery Unit
San Carlo di Nancy Hospital, Rome, Italy

# Foreword to the English Edition

It is with great honour that I received my dear friend and colleague's request to write a foreword for this book. I found it to be a well-formed guide for understanding a multifaceted subject, converging into highly specific therapeutic skills. Only an expert in the subject can convey a clear and accessible message, as this is part of the art of defining an excellent physician. As medicine has advanced and theoretical models have evolved, so too has our understanding of the pathological aspects of disease. The importance of a trustworthy rapport between doctors and patients remains a constant in a civilized society. No artificial intelligence could succeed in such a delicate task without the guidance of an empathetic and enlightened mind, as this book exemplifies.

Laura Giacomelli
Department of General and Specialized Surgery
Sapienza University of Rome
Rome, Italy

"Sapienza" University, Policlinico Umberto I Hospital
Rome, Italy

# Preface

The thyroid is a very important gland, both for the functions it performs under normal conditions and for the frequency of the pathologies that can affect it. Having a thyroid nodule (or even more than one) is far from uncommon, as it can be present in 60–70% of people. Hashimoto's thyroiditis is the leading cause of thyroid dysfunction in areas with sufficient iodine intake, while hyperthyroidism is much less represented epidemiologically, reaching a prevalence of about 1% of the general population. It is also estimated that the prevalence of thyroid dysfunction during pregnancy can range from 0.2% for hyperthyroidism (with severe and frequent complications for the foetus or newborn) to 2.5% for hypothyroidism, up to 4–5% for thyroid nodules.

In light of these simple facts, it becomes evident that thyroid pathologies, due to their widespread occurrence, take on a clear characteristic of a social disease and, as such, should be addressed with the promotion of prevention paths and awareness activities, particularly aimed at ensuring continuous improvement in the quality of life. In this process, which is far from being quick and easy, a key role is played by medical-scientific information.

This is the main reason that impelled me to write this book, a rather "unusual" text— as I like to define it—designed to bring the broad topic of the thyroid closer to those who are not involved in the field. It is structured as a journey into the knowledge of our "butterfly of metabolism". I have tried to explain various pathologies in a simple and accessible language for all sorts of readers, as far away as possible from the technicalities of the medical profession, simplifying diagnostic protocols and therapeutic solutions without prejudice. The only goal is to share information to improve everyone's knowledge and, in the case of illness, provide suitable tools to face it consciously and as calmly as possible.

The idea of integrating text with illustrations revolves around this concept: to shorten distances even further and leave room for the so-called photographic memory, an approach to acquiring basic knowledge through associations of ideas. Drawing, the primary form of communication for every individual, helps to remember knowledge and stimulates awareness of the problem without developing anxiety. In short, this book is for everyone: the young, adults, and the elderly. It is for you, who want to know or delve into the functioning of the thyroid; it is for you who want to learn more to start a prevention journey; it is for you as a thyroid patient who needs to fully understand to accept the disease you are experiencing and make room for a virtuous healing path. And it is for you who care about your well-being and that of your loved ones.

Life is a precious asset, preserve it!

Daniela Pace

# Acknowledgments

Writing this text, in addition to being an opportunity for communication, represented a great challenge for someone like me, who is used to dealing with and expressing herself every day in a technical language, as it is the medical-scientific one. Moving away from technicalities and simply explaining medical concepts was not easy at all: if I succeeded in doing so, thanks to the journalists Roberta Mastruzzi, who supported me in the first phase of drafting, and Fabrizio Giona, who took care of the editing of the book.

I mainly thank them for making it possible to explain topics of physiology, medical pathology, and diagnostic and therapeutic protocols in a language accessible to any reader.

The great work of simplification was embellished by the work of the illustrator Elisabetta Colagrande, who contributed to lightening concepts sometimes difficult to understand, making them more immediate and more comprehensible and to medical interpreter and translator, Ms. Angelica Cicinelli, who revised and proofread the machine translation to ensure that the terminology, content, messaging, and tone of voice of the original were maintained.

A warm "thank you" goes to all my patients because they have been a source of inspiration: it is precisely from their desire to "ask questions," to know and understand better, that the idea of writing a text that could not only explain but also raise awareness, to prevent and take care of oneself in the best way possible.

Last but not least, I thank my family and all the people who love me, and encourage me, supporting me in various professional projects.

# Contents

# Author and Collaborators

## About the Author

**Daniela Pace**  Medical Doctor and Surgeon, specializing in Endocrinology and Metabolic Diseases, and an expert in Neck Ultrasound. She holds a Ph.D. in endocrine, metabolic, and andrological sciences and has a Level II Master's degree in Clinical and Interventional Ultrasound of the neck. Currently working at the San Carlo di Nancy Hospital in Rome, she is also a teacher at the SIUMB School (Italian Society for Ultrasound in Medicine and Biology) for ultrasound diagnosis and percutaneous therapies of endocrine pathologies—"Save your thyroid" International School and Training Lab of interventional ultrasound-assisted procedures.

## Collaborators

**Roberta Mastruzzi**  Journalist and publicist. She has collaborated with numerous magazines in the health, wellness, and nutrition sectors. She is the author of the fiction book for children "Quanti pasticci, Ricottina" (Einaudi Ragazzi, 2017). She is currently engaged in journalistic activities and public relations.

**Fabrizio Giona**  Journalist and publicist. He has collaborated with newspapers, magazines, and information portals and has published scientific works on health and safety at work. Currently involved in editorial projects and cultural promotion, he also handles press office and digital communication for events, personalities, and public administration.

**Elisabetta Colagrande**  Art teacher, painter, anatomical illustrator, and cartoonist. Featured in numerous art exhibitions with her oil paintings. She creates drawings and anatomical illustrations for various medical-scientific books (CT and MRI of the knee, MRI of the biliary tract, Imaging of the shoulder). Recently, she has been the creator of comic vignettes for medical texts (*The Parkinsonian Patient*, ELI Medica 2020).

# All About Your Thyroid

**1**

**All About Your Thyroid**

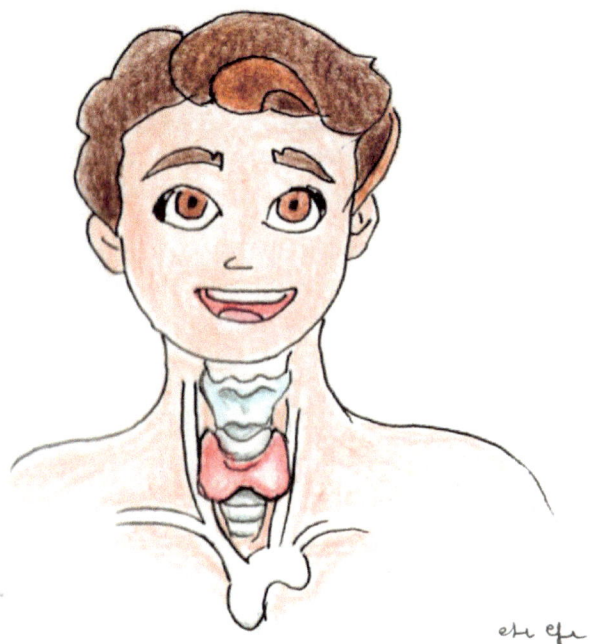

The thyroid is a small, yet very important gland for our well-being as it produces and releases hormones that are essential for functioning of our body. These hormones are small and fast messengers that each time the body commands, dive into the blood vessels and spread throughout our system, carried by the blood flow. Once inside the cells, which our organs are made of, they release valuable information that allows the organs to work at their best.

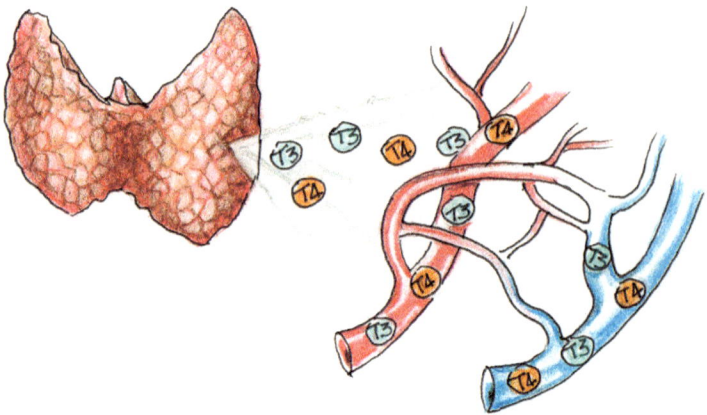

Do you have any idea how many functions our thyroid controls? We must imagine it as the conductor of a large orchestra, who—to ensure the audience of a great performance of harmonious music—must coordinate all the musical instruments. In the same way, our gland is called to regulate the organs activities, allowing the body to stay healthy and balanced. However, some things may not always work as they should.

For instance, it can sometimes happen that the thyroid function slows down or accelerates: if that's the case, we must intervene as soon as possible to help our *conductor* perform perfectly! How can we do it? Let's take a look together, step by step.

## 1.1   Thyroid: Anatomy

The thyroid is located in the lower part of the neck, just above the *jugular notch* or *suprasternal notch* (a large depression on the top of the sternum) and just below the *Adam's apple* (the cartilage that covers the front of your larynx). Its shape is very similar to that of a butterfly, with two large lateral wings, called **lobes**, joined together by a thin strip of tissue called **isthmus**. In adults, each lobe has an average length of 4–5 cm, a width of 1.5–2 cm, and 1–1.5 cm thick. The thyroid, moreover, is a real "featherweight" with its 20–25 g!

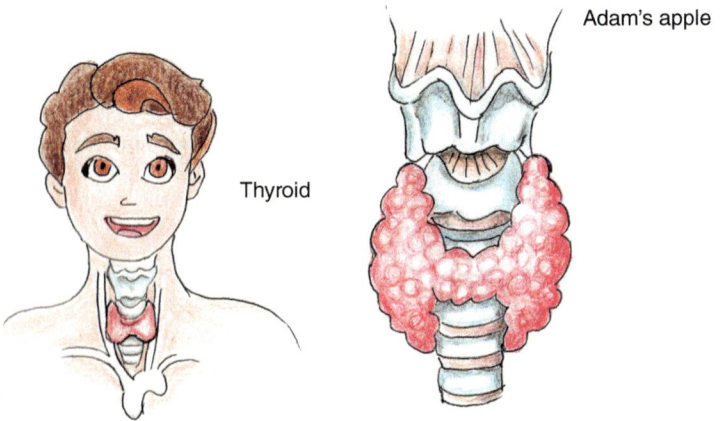

Adam's apple

Thyroid

The lobes are placed around the trachea, a small tube that carries air from the throat to the lungs as we breathe, and around the larynx, the area in the throat housing the vocal cords, allowing us to make sounds. If we could touch it, we would discover that the surface of the thyroid is not smooth, but composed of many small spheres-like vesicles, called **follicles.** The follicle is the functional unit of the thyroid, and its structure is made up of a layer of cells, called **thyrocytes**, which delimit a cavity. These are connected to our blood system by many thin capillaries. Inside the follicles, there is a very dense liquid, composed of a gelatinous substance called **colloid**, and this is where the synthesis of **thyroid hormones** takes place. At the same time, the colloid acts as a storage, where it preserves and maintains the hormones *ready to go*. Between one follicle and another, there are parafollicular cells, which we can abbreviate as **C cells**: C as in Calcitonin, the hormone that regulates calcium metabolism. We will soon see what this means.

## 1.2 How It Works

The main function of the thyroid is to release a steady amount of thyroid hormones. These have a skeleton, a framework called **tyrosine**, to which atoms of **iodine** are attached. When the iodine atoms are 3, the hormone will be nicknamed **T3**. When, instead, there are 4 of them, it will be nicknamed **T4**. Sometimes, someone calls them by their given name: *tri-iodothyronine* (T3) and *tetra-iodothyronine* (T4), abbreviated as *thyroxine*. However, the thyroid does not autonomously choose to produce hormones. The one giving the command is another gland, the **pituitary gland**, a small outgrowth, pea-sized gland located in the brain, at the base of the head. The pituitary gland controls the activity of most other hormone-secreting glands, but the one that involves our butterfly is called TSH, which has the task of communicating the amount of T3 and T4 to produce.

Once generated, these hormones persuade the pituitary gland to reduce TSH production.

It is clear, from what has just been said, the indispensable and crucial role of the pituitary gland: it is the *thyroid controller*, always attentive in checking that it performs well and therefore that the hormones are in circulation. When it perceives that our gland relaxes too much, it sends TSH again, to stimulate the thyroid to work again. This mechanism of reciprocal responses, called *feedback*, is thus set in motion.

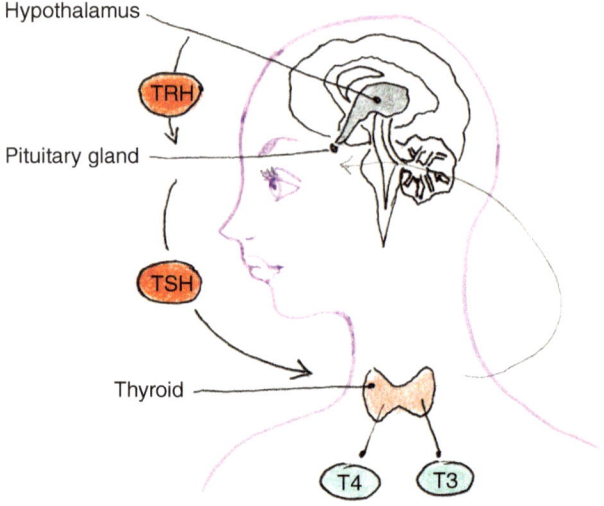

Once the thyroid has produced its T3 and T4 hormones, part of these are bound to a protein that, like a small raft, transports them in the blood through the blood vessels, accompanying them on their journey along the entire human body. But it is the hormones unbound from the raft-protein, the ones free to swim in the bloodstream, that manage to perform their messenger activity. Since they own this freedom, they are called **Free** and abbreviated as **FT3** and **FT4**.

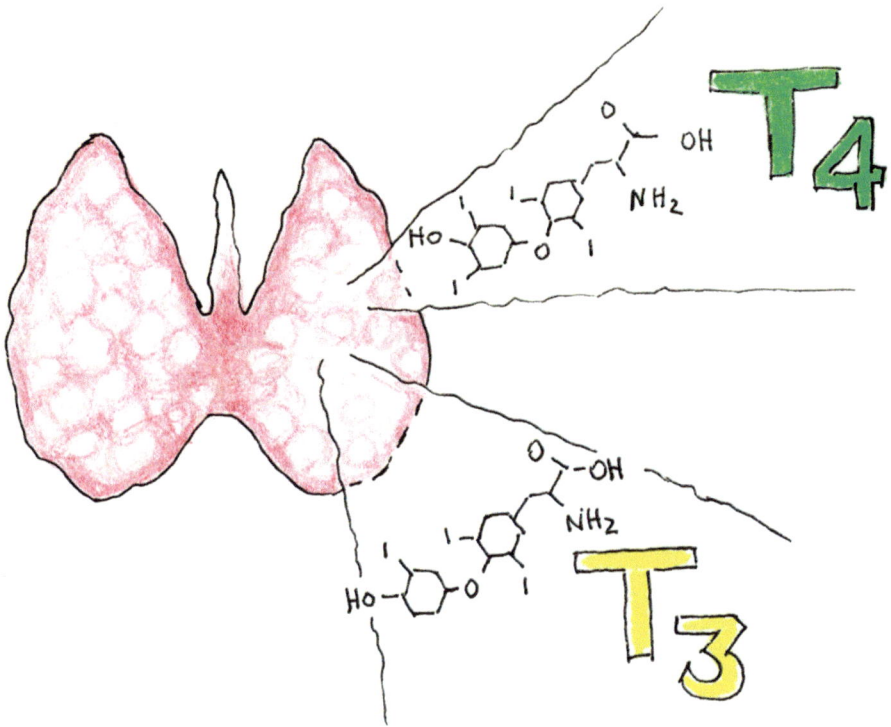

Finally, through the **C cells**, the thyroid produces another hormone, called **Calcitonin**, which has the task of conveying to bones, intestines, and kidneys the important message of maintaining the right presence of calcium, stimulating its reduction when it is too high.

## 1.3    What We Need It For

Through its hormones, the thyroid performs numerous functions: it intervenes on bones, muscles, teeth, and hair growth and development; it acts on food metabolism, specifically in the way in which we manage to transform the food we eat into energy, especially fats and sugars; it regulates the frequency of the heartbeat, body temperature, and many activities of our nervous system, including interfering with our mood; moreover, it affects your skin, vision, menstrual cycle, and the efficiency of our mental abilities, such as memory and concentration.

Its correct functioning makes us feel good: we are full of energy and eager to do anything, we feel fit and lively, ready to engage in our daily activities, play sports, and have fun with our friends. On the contrary, if we are often tired, exhausted, nervous, weighed down, lazy, and demotivated, there might be something in the laborious mechanism of thyroid hormone production that is not going in the right way. That's when the **Endocrinologist**, the doctor who specializes in the diagnosis and treatment of the thyroid (and not only that!), gets to work like a detective to investigate the cause of its malfunction.

## 1.4    Thyroid Best Friends: Iodine and Selenium

The **iodine** is an essential mineral for the T3 and T4 hormones production, as we have already seen, it is part of them. Our body absorbs it from food and water.

Iodine comes from the sea: that's where it's born, then when it evaporates, it condenses forming the clouds turning into water droplets (rain) or ice crystals (snow), thus arriving—transported by raindrops or snowflakes—to distribute itself all over the land surface, from the coast to the mountains.

From here, some water penetrates the soil and it is absorbed by plants or animals, and finally by us—human beings—when we eat or drink.

Thyrocytes take iodine from our blood and carry the mineral to the thyroid so that it can be used for hormone construction.

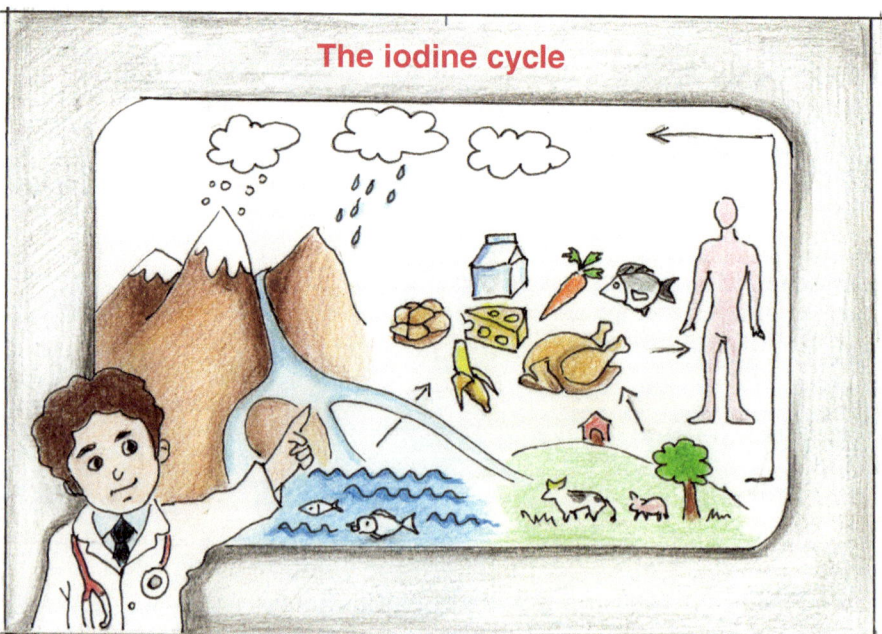

### If Iodine Is Lacking

It often occurs that the iodine concentration in our body is not sufficient. This may happen for several reasons, and first and foremost because its distribution is not homogeneous (e.g., iodine deficiency occurs in those areas very far from the sea,

**Table 1.1** Consequences of iodine deficiency

| In all ages | Goiter |
|---|---|
| In the fetus and newborn | Abortion and perinatal mortality, cretinism, hypothyroidism psychomotor defects |
| In children and adolescents | Subclinical and overt hypothyroidism, psychosomatic development delay |
| Adult/elderly | Nodular goiter, hyperthyroidism from nodular goiter |

*Consequences of iodine deficiency, modified from "ICCIDD, UNICEF, WHO 2001 Assessment of Iodine Deficiency Disorders and Monitoring their Elimination"*

because the plants that grow there and the animals that feed on them do not contain enough); moreover, our daily diet may not include iodine-rich foods: as you can imagine, real treasure chests containing precious treasures of iodine are the creatures that come from the sea, therefore fish, seaweed, crustaceans, and mollusks.

We can check the iodine deficiency in our body by measuring the quantity in an urine sample, in the laboratory: the value that interests us is called **ioduria**.

What happens if we were to discover that the iodine level is too low? We have seen that the pituitary gland, through the TSH hormone, stimulates the thyroid to produce T3 and T4 and, at the same time, it brings it to a halt when the amount of hormones in circulation is enough, through a feedback mechanism resembling that of a radiator thermostat: we set the temperature we want in a room and when it is reached, the system stops, only to restart in case the temperature would decrease or increase.

If the thyroid, due to lack of raw material—iodine indeed—cannot produce the required amount of hormones, it will be continuously stimulated by the pituitary gland to produce them, without ever interrupting its work, thus leading to overexertion. In a few words "it gets stressed," as much as a person is forced to work non-stop, not being able to recover. Excessive overexertion can over time lead to the development of various disorders, such as the enlargement of the thyroid (**goiter**) or to dysfunctions like **hypothyroidism** or, in newborns, it can cause a developmental delay of their mental abilities as well as the muscle and bone development (Table 1.1).

## A Pinch of Intelligence: Use a Little Bit of Iodized Salt!

It is important to ensure our bodies have plenty of this mineral. My advice is to take a small amount every day, and the easiest way to do so it's through diet.

Among the most iodine-rich foods, we find fish and shellfish, which we do not eat every day, as well as beans, zucchini, and spinach. Very often, however, these are not enough to guarantee the daily requirement of iodine.

So, what to do? The answer is simple. Just use a pinch of **iodized salt** when cooking, a type of salt that contains the dose of iodine our body needs every day.

If you think about it, we use salt daily when preparing pasta, seasoning vegetables, and as a meat condiment. Therefore, replacing the traditional type with the

| **Table 1.2** Recommended iodine intake (mg/day) | Children 0–5 years | 90 |
| --- | --- | --- |
| | Children 6–12 years | 120 |
| | Adults over 12 years | 150 |
| | Pregnancy | 250 |
| | Breastfeeding | 250 |

*Recommended iodine intake (mg/day), modified from "ICCIDD, UNICEF, WHO 2001 Assessment of Iodine Deficiency Disorders and Monitoring their Elimination"*

iodized one (be careful, one must not exaggerate with the quantities!) can certainly help us in the mission of providing our thyroid with the raw material it is so "greedy" for.

One can easily find it at the grocery store and supermarkets and don't worry, the word *IODIZED* is always clearly printed next to the word *SALT*.

When should you use it? We should start from childhood, then continue as teenagers as well as in adulthood, and even when we get older.

Every stage of life requires the right dose of iodine! The intake during pregnancy, it's really important, since it is precisely throughout this phase that the child's nervous, muscular, and skeletal system is developed (Table 1.2).

## Selenium, a Great Ally

Another fundamental mineral for the well-being of our thyroid is **Selenium**. Its name comes from the word *Selene* which means *Moon* in Greek: its silvery color reminds of the lunar surface! Discovered about 200 years ago, by a Swedish chemist, Dr. Jöns Jacob Berzelius, it has been defined as as an essential micronutrient: this means that its presence is essential for the entire human body to function. In the same way as iodine, we get selenium through diet, and the mineral plays a critical role in many bodily functions, including the thyroid. Indeed, it supports our gland in the production of hormones T3 and T4 and, at the same time, acts as a sweeper, as it eliminates the waste that is created in the production process, namely the molecules of **hydrogen peroxide** (code name $H^2O^2$). Broadly speaking, selenium slows down the aging process of all our cells, keeping them always active and efficient: this particular ability is called **antioxidant function**.

We find this mineral in fish and shellfish, meat, milk, and yogurt, flour-based products such as pasta, rice, and bread, many vegetables like cabbage, broccoli, cucumbers, garlic, and onion, as well as in dried fruit, especially in one with almost "magical" powers, so-called *Brazil nut*. This is the food highest in selenium! One or two of these nuts a day would be enough to ensure our daily intake.

Even though selenium is present in products that we consume almost every day, like pasta and bread, we must take into account that a part of the mineral is lost due to long cooking times at high temperatures. Along the line of iodine, its distribution is not homogeneous: some soils are richer than others. Research shows that we Europeans take a slightly less amount of selenium than we need since some modern cultivation methods deplete the soil.

## The Winning Duo!

Iodine and selenium represent a winning combination: they are the antidote to keep our thyroid in shape.

It's not surprising that the Ministry of Health promotes the consumption of these two micronutrients; the recommended intake for adults is 150 μg of iodine and 40 μg of selenium. Still, the doses greatly vary depending on age and the stage of life (e.g., when a mother has a "bun in the oven" and when she's breastfeeding, she should increase the amount of iodine up to 150–200 μg).

Following the principle that *Prevention is better than cure*, Italy has issued a Law (no. 55 of 2005) in order to promote the so-called **Iodoprophylaxis**.

A complicated word to pronounce, yet easy to put into practice: as a matter of fact, it encourages families to use iodized salt instead of the common one. A simple but effective action, which can keep us away from many diseases!

## 1.5      Parathyroids, the Calcium Guardians!

Before we carry on the thyroid, I want to introduce four of its friends, called **parathyroids**.

They are glands too, much smaller, and they live next to our butterfly (to be precise they are lying on its wings): two of them are located behind the right lobe, one on top and the other at the bottom; the other pair is the specular reflection, behind the left one.

Slightly larger than a lentil, these glands are fundamental for our well-being as they are responsible for the production of **parathyroid hormone (PTH)**, the hormone that regulates the presence of phosphorus, the activity of vitamin D and, above all, the **Serum calcium**, that measures the amount of **calcium** in your blood as well as the one stored in the skeleton.

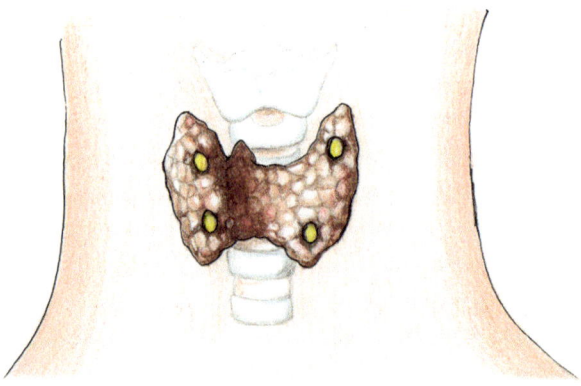

## How Do They Work?

One should know that calcium is essential for our well-being. Did you know, for example, that our bones are composed of 90% of calcium? This precious mineral is vital for the transmission of impulses of the nervous system, muscle contraction, and coagulation. In order to work properly and make us feel good; however, it must be present in just the right amount: not too much, nor too little. And right here the parathyroids come into play!

Let's see how they work. When the presence of calcium lowers in our blood, the parathyroids begin to produce greater amounts of PTH, in a perfectly synchronized mechanism. This stimulates the bones to release part of the calcium they had stored in their deposit, just like a safe deposit box at the bank. In addition, the parathyroid hormone, to increase the calcium levels in the bloodstream, stimulates the production of vitamin D from our kidneys, which in return, help our intestinal absorption as much as possible, from the food we eat. Lastly, it will demand the kidneys to filter and retain calcium so that it won't be excreted in the urine; this way, there will be more in circulation.

Thanks to all these measures taken by the PTH, when the calcium levels return to normal, the parathyroids can finally sigh in relief and stop the extra production of the hormone.

## And If They Lose Balance?

Sometimes, this wonderful mechanism may not work as it should. The parathyroids might start to work too much or too little. In the first case, we identify **hyperparathyroidism**: the glands produce too much PTH hormone which leads to an excessive release of calcium from the bones. As a result, we might experience some disorders, related to the loss of calcium from bones, as well as an excess of this mineral in the circulatory system: so we might suffer from weak, and fragile bones or even from osteoporosis (a condition that develops when the bone mineral density and bone mass decrease, leading to the increasing risk of broken bones at the slightest trauma), along with chronic fatigue, muscle fatigue, poor concentration, kidney stones, abdominal pain, and so on. Some drugs can diminish all these symptoms, although the definitive solution would be the surgical removal of the parathyroids.

On the other hand, if these glands work too little, **hypoparathyroidism** will be diagnosed: the low production of PTH leads to poor calcium concentration along with high phosphorus levels in the blood. This disparity can lead to disturbances such as annoying tingling or numbness, up to muscle cramps, and other consequences.

In the second case, the treatment involves dietary supplement intake of larger amounts of calcium and vitamin D, and if needed, drugs containing these two elements, as a way to compensate for those which are not produced by the parathyroids.

One can check it with a simple blood test, which inspects the values of PTH, vitamin D, and calcium in our bloodstream (in the urine too), an endocrinologist then can have a better picture of the current situation to understand if these small glands, so useful for our well-being, are doing their job well! If this is not the case, the physician will work to understand what has hindered their regular functioning and if behind an excessive or poor work of the parathyroids, there may be the influence of some other disease.

**Remember That…**

√ The thyroid is a small butterfly-shaped gland located at the front base of the neck. It regulates many functions in our body through the production of thyroid hormones called tri-iodothyronine (T3) and thyroxine (T4). The production of these hormones is, in turn, controlled by the hormone TSH, which is produced by the pituitary gland, a tiny gland located inside our skull.

√ Through its hormones, the thyroid performs numerous and important functions to ensure our well-being: it plays a role in growth, acts in the metabolism of food, regulates heart rate, body temperature, and many activities of our nervous system. It also influences respiration, skin health, vision, the menstrual cycle, and the efficiency of our mental capacities, such as memory and concentration.

√ Its regular functioning is ensured by an adequate intake of iodine, an essential mineral for the production of thyroid hormones (T3 and T4). It is present in the human body in quantities of 15–20 mg, and the estimated daily requirement is 150 μg. However, we often struggle to consume the right daily doses of iodine because its presence in food and water varies significantly. That's why it is essential to use "iodized salt" in the kitchen.

V Iodine deficiency is considered by the World Health Organization as one of the most serious public health problems. The shortage of this mineral in our body can lead to various pathologies, more or less severe depending on age and gender, such as hypothyroidism or hyperthyroidism. The need for iodine is particularly high for pregnant women and children.

V Another ally of our thyroid is undoubtedly selenium, which serves as an anti-oxidant. Like iodine, selenium is present in many foods, but its concentration varies depending on the region. I recommend eating two Brazil nuts a day, as they are the food that contains the highest amount of selenium!

# When the Thyroid Gets Sick

**2**

**When the Thyroid Gets Sick**

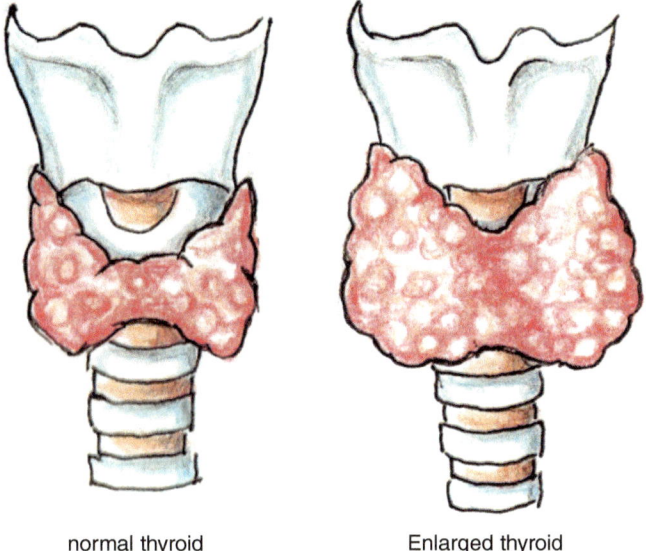

normal thyroid          Enlarged thyroid

As anticipated in the previous chapter, the thyroid can get sick. Among the most common problems there are goiter, hypothyroidism, hyperthyroidism, thyroiditis, and nodules. In the following pages, you will learn how to recognize them, as well as to prevent them or, in case you have already received a diagnosis, to deal with the treatment. In any case, do not panic because we have a solution for many of these diseases!

## 2.1    Goiter

Iodine plays a really important role in the well-being of the thyroid. Its deficiency does not go unnoticed as it manifests itself with the enlargement of our gland, hence one may notice a lump, or swelling in the lower part of the neck, which becomes even more evident if we recline the head back. After a careful palpation, the doctor will evaluate if our butterfly "has put on" a few extra grams and they will advise us to run some simple tests, such as ultrasound and blood work. If the suspicion is confirmed, then we will delve into the disorder we suffer from: a **goiter**.

### What Is a Goiter?

The pituitary gland, through its messenger TSH, stimulates the thyroid to produce the hormones T3 and T4, do you remember? That's why, if there was not enough iodine, our gland would not be able to produce enough hormones and, as a result, it would be continuously stimulated to work, right? Well, this work overload can cause the enlargement of the follicles that form the thyroid tissue, thus appearing larger than usual: in this case, we are in the presence of a **diffuse, or simple, goiter**.

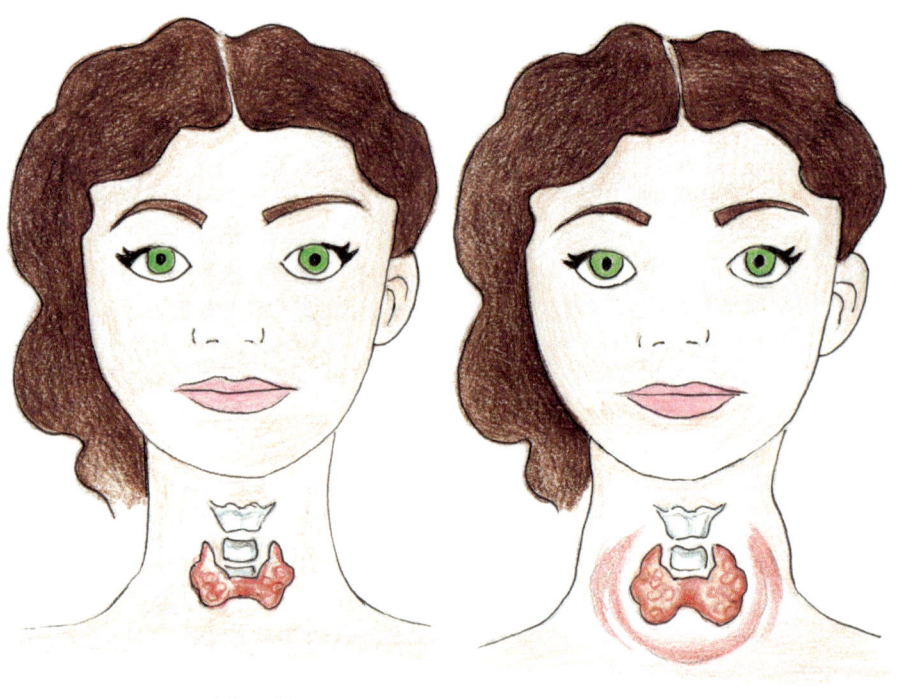

normal thyroid                                              goiter

However, the increase in volume does not affect the entire gland sometimes, but it could be restricted to a specific area and accompanied by the presence of one or more nodules, respectively called **uninodular goiter (a) and multinodular goiter (b)**.

It is good to know that usually—even in the presence of a goiter—the thyroid manages to perform its functions normally; in this case, it is defined as a **non-toxic goiter**.

Despite the benignity, it is always good to keep an eye on it, because it could cause discomfort when we breathe, speak, or eat.

Since the thyroid lies on the trachea, its increase in volume could compress it, causing the above-mentioned problems. It should be constantly monitored over time since the nodular goiter could become toxic; the thyroid begins to produce T3 and T4 independently, without listening to what the pituitary communicates through the TSH. As a result, we will find little TSH and a lot of T3 and T4 in the blood tests. In this case, we'll have a **toxic multinodular goiter**.

This thyroid **functional autonomy** is not a good thing, since it could lead to other disorders in the long run.

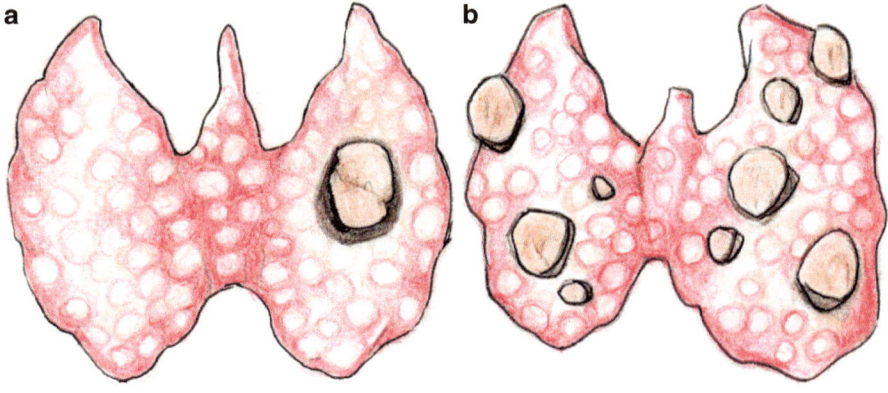

## What Can Be Done to Prevent Its Formation?

Taking into consideration that the most common reason for the development of a goiter is the iodine deficiency, the first thing to do is to use iodized salt. A small amount, every day, is enough to restore the right levels in our body, thus allowing the system to return to a perfect balance between work and rest, production and relaxation!

## 2.2    If the Thyroid Works Too Much... or Too Little

Our thyroid is exposed to two kinds of risk: working too much, like a tireless and hardworking ant, or basking in idleness, until it becomes as lazy as a lizard in the sun. In both cases, something is wrong; the perfect balance between production and rest becomes unstable, causing the scale to tip to the one side or the other. This is where we need to be alert and start investigating, so we can understand if our thyroid is getting sick. I will now explain **hyperthyroidism** and **hypothyroidism**, two terms indicating thyroid dysfunctions, which are responsible for various disorders and diseases that can affect the gland, such as goiter, thyroiditis, and Graves-Basedow disease.

### An Overactive Thyroid

As we know, all words that start with *hyper-* indicate something excessive or taken to the extreme. Therefore, when we hear the doctor talking about **hyperthyroidism**, it means that the thyroid is working too much. This huge and constant workload is not a good thing since it pushes our gland to produce higher numbers of T3 and T4 hormones. As a consequence, in our body, there'll be a lower amount of TSH, the pituitary messenger hormone. We can appreciate this with a simple blood test, and by paying attention to our body's reactions. Let's see some simple clues.

- **Weight**. Excessive and sudden weight loss can be the first symptom: this is because thyroid hormones affect our metabolism which, in the case of hyperthyroidism, turns out to be accelerated. In short, the energy consumption is higher than we supply.
- **Feeling hot**. T3 and T4 hormones regulate our body temperature; therefore, their exertion could increase the body temperature, developing greater heat intolerance and warm and clammy skin.
- **Heart**. The heartbeat is sometimes a bit too fast as if we had just finished a run. This symptom comes from the fact that thyroid hormones affect the proper functioning of our cardiovascular system too, dictating the rhythm at which the heart beats, along with the volume of blood poured into our blood vessels, which, if excessive, can lead to **high blood pressure** (when the pressure in your blood vessels is too high) in the long run.
- **Stomach ache**. In the case of hyperthyroidism, we might suffer from episodes of diarrhea or irritable bowel, because the transit of food in the intestine is accelerated. Other frequent alterations of stomach and intestine function may be abdominal pain, vomiting, and **dysphagia**, literally, the difficulty in *swallowing* food!
- **Menstrual cycle**. Girls and women with hyperthyroidism can have very close periods, with breaks between one menstruation and the next one shorter than the canonical 28 days.

- **Mood disorders**. Memory and poor concentration, maybe poor academic performance, mood swings, nervousness, insomnia, and a sense of restlessness can all be signs of the excess of thyroid hormones in circulation.

On a side note, these symptoms are not always present, in fact, some of them, never show at all (e.g., some people do not lose weight, even though they suffer from hyperthyroidism), while others are very subtle or could only be occasional and have nothing to do with the thyroid well-being. In any case, it is always good to talk about it with your family doctor, who will be able to put together all the clues and start to investigate, running specific tests.

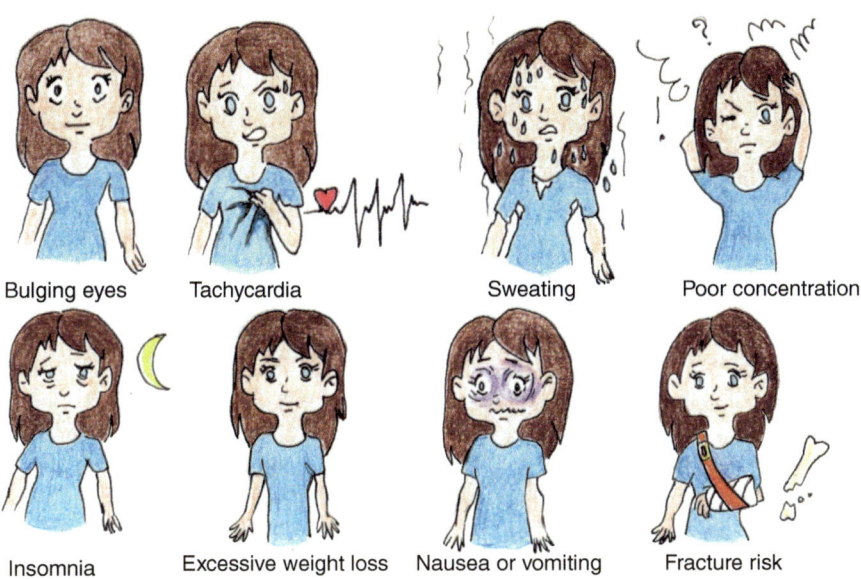

Bulging eyes    Tachycardia                    Sweating         Poor concentration

Insomnia    Excessive weight loss    Nausea or vomiting    Fracture risk

## Why Does It Happen?

Why has the thyroid suddenly started working so hard? The causes of hyperthyroidism can be different. In most cases, it could be the evolution of an **uninodular goiter** or **multinodular** that we have neglected: one of the nodules present in the thyroid begins to function more than it should, producing excess thyroid hormones.

Either it could be the consequence of a disease, called **Graves' disease** where the gland appears enlarged, as well as swollen and protruding eyes as if they were always wide open (as an obvious external sign). Rarely this could be the consequence of **thyroiditis**, or an excess of iodine in the diet, it may also be related to pregnancy or even determined by other factors.

It will be the endocrinologist's job to find out what caused this overexertion and prescribe the most suitable treatment to bring everything back to normal.

## How Is It Treated?

Some kinds of drugs are used to treat hyperthyroidism and can block the production of new hormones. These are called *antithyroid drugs* or *thyrostatics*.

Like all medicines, these can bring *side effects* too, unwanted and even dangerous reactions so to speak, that sometimes appear when a substance used to treat an organ is not properly "digested," thus causing discomfort or even damage to other organs.

This is the primary reason why medicines must always be prescribed by our family doctor, or a specialist, since they know how to dose them, calculate their risks and benefits, and above all, they know which ones are most suitable for us, depending on our current health and our medical history. Among the undesired effects of *antithyroid drugs*, there is leukopenia (low white blood cell count, a decrease in their number), hepatotoxicity (impairment of the liver function, which could be a chemical-driven liver damage), and itching.

There are other types of therapies used to treat hyperthyroidism. To drastically stop the thyroid activity, it is possible to plan the removal of the thyroid, or just part of it, through surgery or radioactive iodine, destroying the gland tissues.

These cures, however, have their side effects too! In fact, the thyroid—either completely or partially removed, or destroyed by radiation—may risk becoming incapable of producing the right amount of hormones, falling exactly into the opposite situation, hypothyroidism.

## A Lazy Thyroid

**Hypothyroidism** occurs when the thyroid is a bit too lazy and our butterfly is very slow in performing its job. As a result, all the organs slow down their functions and so we may put on a few extra pounds, we may often feel tired and with little desire to be active.

Let's take a look in detail. If you have carefully read all the clues that make us suspect hyperthyroidism, you will not find it difficult to memorize those that, instead, lead us on the path of hypothyroidism, because they are exactly the opposite!

Here, the T3 and T4 hormones in circulation are fewer, while there is a lot of TSH, sent from the pituitary gland to the thyroid in an attempt to persuade it to work more.

- **Weight**. Stepping on the scale we immediately notice some weight gain, due to a slowed metabolism, where we do not manage to transform what we eat into energy so the excess turns into fat. In addition, there will be a greater stagnation of fluids, so looking at the mirror we will see ourselves as more swollen, especially in arms and legs.
- **Cold sensitivity**. In the case of hypothyroidism, we tend to become "cold-intolerant," so we feel more discomfort in cool environments than those around us.
- **Heart**. The heartbeat is slowed down, and there may be repercussions on the cardio-circulatory system too: fewer and weaker heart contractions can lead to different consequences, such as the accumulation of cholesterol and other deposits on the artery walls.
- **Stomach ache**. We talk about **constipation** when the appointment with the stool is postponed day by day (in some cases, it may even take days!). This happens because the digestive system slows down and quite often hypothyroidism is associated with **gastroesophageal reflux disease (GERD)**, a condition that occurs when the food we eat "comes back up" (the stomach acid repeatedly flows back to the esophagus) disturbing our digestion.
- **Menstrual cycle**. Girls and women with hypothyroidism have short and irregular periods, it may occur that some months, the cycle may even be completely absent!
- **Mood disorders**. We often feel tired and inactive and just like in hyperthyroidism, there may be problems with memory and concentration along with mood swings. Still, the most common feeling is certainly laziness!
- **In the mirror**. Our skin is very dry, we tend to have thin and fragile hair, and nails break easily and do not grow.

Even in this case, the clues may not all show up together and, especially at the beginning, may not be very clear.

**Subclinical hypothyroidism** occurs when symptoms are mild or almost unnoticeable, characterized by high levels of TSH yet normal values of T3 and T4. While **clinical hypothyroidism** occurs when there is a reduction of these two hormones; in this form, the symptoms are more evident.

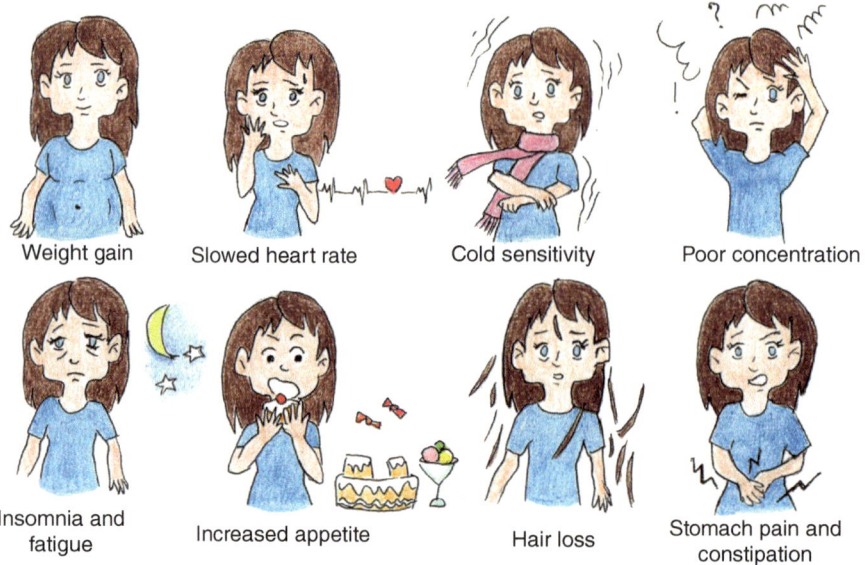

Weight gain          Slowed heart rate          Cold sensitivity          Poor concentration

Insomnia and          Increased appetite          Hair loss          Stomach pain and
fatigue                                                                constipation

## Why Does It Happen?

Some people suffer from hypothyroidism from birth, others are affected in adult-hood. In some cases, as we have said, this condition is a consequence of the surgical removal of the thyroid, or part of it, or a therapy with radioactive iodine. Beyond these situations, the thyroid can also function poorly if attacked by bacteria (**acute thyroiditis**) or viruses (**subacute thyroiditis**) or, still, if affected by **Hashimoto's thyroiditis**, an *autoimmune* disease. Its meaning indicates a state in which our immune system—the one that sends its soldier-antibodies to protect us, for exam-ple, from the flu—targets the wrong enemy and sends its warriors to fight against internal organs, rather than against external ones, like viruses and bacteria. But we will talk better about this shortly.

## How Is It Treated?

If we are early diagnosed with hypothyroidism, we can soon help our thyroid to start working at the right pace again, just by the right amount of iodine and selenium intake. In other cases, however, the doctor will prescribe a drug that helps us to cor-rect the deficiency of T3 and T4 hormones (since our gland is unable to produce enough), thus restoring normality. This drug is called **levothyroxine**, a synthetic version of a hormone that replaces the T4 deficiency, performing the same func-tions. On the other hand, this is not enough to defeat hypothyroidism: first of all, we must take care of ourselves, and improve our lifestyle. This translates into introducing the right amount of essential nutrients for the thyroid, such as selenium, iodine, and zinc, a balanced diet, physical activity, and learning how to manage stressful situations—that often push us to eat even when we're not hungry- and sleeping at least 7 hs every night. These are golden rules for everyone, especially for those who suffer from a thyroid disorder.

## 2.3  Thyroiditis and Autoimmune Diseases

As we have seen, **hyperthyroidism** and **hypothyroidism** are the two main manifestations of thyroid diseases. They can be caused by various factors but among them, we undoubtedly find *thyroiditis*, a term that indicates a state of inflammation of our gland. There are several types with high-sounding names, such as **De Quervain's Thyroiditis**, also known as subacute granulomatous thyroiditis, and **Chronic Lymphocytic Thyroiditis**, also known as **Hashimoto's Thyroiditis** or **Hashimoto's disease**. Let's get to know them together!

### Subacute Thyroiditis

The thyroid might get sick because it is hit by a **virus** that causes an infection, which, as a domino effect, causes inflammation of its tissues. This is the same defense mechanism as when we are hit by the flu: our throat appears very red and gives us an annoying burning sensation. De Quervain's **Thyroiditis** occurs when a virus hits the thyroid. Its name originates from the Swiss surgeon who first described it in 1936: Fritz de Quervain.

Subacute thyroiditis

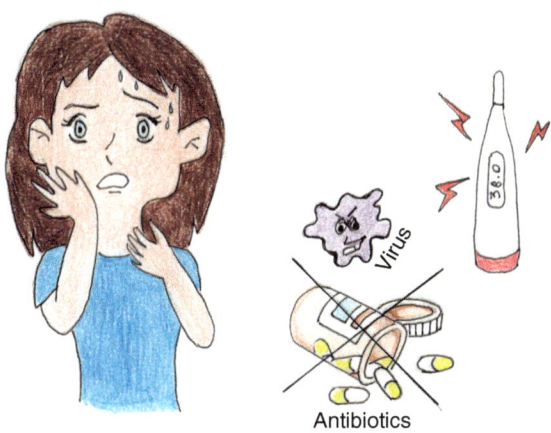

The most common symptoms are moderate-to-severe neck pain and tenderness, Sometimes the pain is extended to different areas, such as the jaw and ears. In some cases, fever and fatigue may occur.

Let's see what happens in our body. The gland, attacked by the virus that temporarily destroys its follicles, reacts releasing all its hormones at first, which thus circulate in the blood manifesting the symptoms linked to hyperthyroidism. Right after this first phase, a period of hypothyroidism is followed, since the gland—malfunctioning—is unable to produce new hormones. It will resume its work only after completely defeating the virus: in some cases, it will take a few months before everything returns to normal. Fortunately, the symptoms of inflammation are transitory!

## Acute Thyroiditis

The enemies of our butterfly are not viruses only. Although it may be rare, a thyroid infection can also be caused by **bacteria**, and this is the **Acute Thyroiditis** case. The symptoms are very similar to the subacute ones: neck pain, fever, and swollen glands. In addition, the skin of the neck, right at the thyroid level, may appear warm and reddened, a sign of ongoing inflammation. If immediately treated with antibiotics, the infection passes quickly without leaving a trace, if neglected though, it can lead to various complications.

That's why it is vital to intervene as soon as possible, relying on your endocrinologist.

Acute thyroiditis

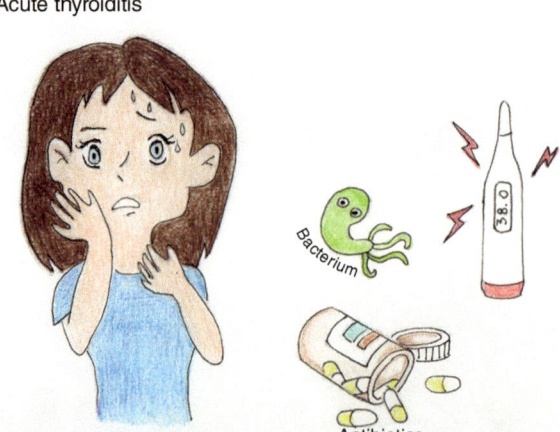

The doctor will be the one to distinguish the two types of infection and to evaluate the treatment to adopt: if the origin is a virus, antibiotic treatment won't most likely produce any effect, since this type of medicine is effective only against bacteria. This is a very important rule to keep in mind for any type of infection: viruses and bacteria are different enemies and need to be fought with different weapons!

## Autoimmune Diseases: Hashimoto's Chronic Thyroiditis

Hypothyroidism may originate from a disease called **Hashimoto's Thyroiditis**. In 1912, a Japanese doctor, for instance Dr. Hashimoto, was the first one to describe the condition and only in 1956 has been discovered to have a particular characteristic: being an autoimmune disease.

The Immune system

Hashimoto's Thyroiditis
(autoimmune disease)

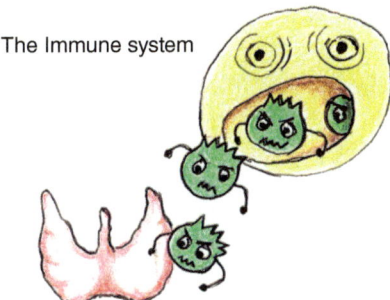

produces lymphocytes against
the thyroid cells

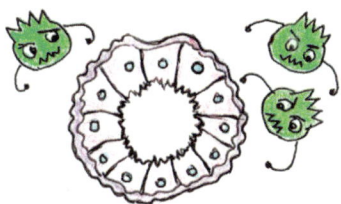

antibodies destroy the thyroid
follicular cells

= Inflammation of your thyroid gland

This means that those who are affected by it assemble antibodies and lymphocytes directed against the gland itself: these are called **autoantibodies**.

But let's take a step back and see what it's all about. You should know that, normally, our body's defense system—the so-called immune system—distinguishes foreign substances from cells of our body, producing antibodies only when it perceives that the latter has been exposed to a threat from the outside, like viruses and bacteria.

It can happen, however, that our defense system "goes crazy" and identifies one (or more) of the natural substances of our body as a source of external danger. It will thus start to produce the autoantibodies and use them to destroy its own cells, tissues, or organs, causing inflammation, damage, or dysfunctions that can lead to the appearance of autoimmune diseases. So, in our case, the immune system does not recognize thyroid tissues as part of the body and produces autoantibodies to attack them. These tissues, which used to be perfectly functioning, after the numerous attacks perpetrated, inevitably become inflamed and, over time, so damaged, that are no longer able to produce hormones.

In other words, the predominant phenomenon becomes the production of autoantibodies directed against some components of the thyroid, which causes progressive damage to the gland itself. This can call forth the surfacing of a goiter due to the increase in volume or, as often happens, to hypothyroidism (sometimes goiter and hypothyroidism can also appear together) or, again, to temporary hyperthyroidism, which then gives rise to hypothyroidism.

In other cases, such evident signals may not even show: one can remain in a state of "truce" for years, despite the presence of autoantibodies. In short, it is a puzzle our endocrinologists have to solve! The triggering cause is unknown although there is a genetic predisposition.

Generally, if we discover that we are predisposed to Hashimoto's thyroiditis, we don't have to do anything, just run some periodic examinations to ensure that the thyroid is doing its job well. However, if symptoms start to become evident, the TSH value is above normal or we are planning for pregnancy, the endocrinologist might decide that it's time to intervene with **levothyroxine**, which I have already described in the previous paragraph dedicated to hypothyroidism.

## Graves' Disease

It can sometimes happen that autoantibodies stimulate the thyroid to an excessive production of hormones, causing a condition of hyperthyroidism. In this case, the autoimmune disease is called **Graves' disease**, determined by a predominant production of TSH receptor antibodies (TRAb).

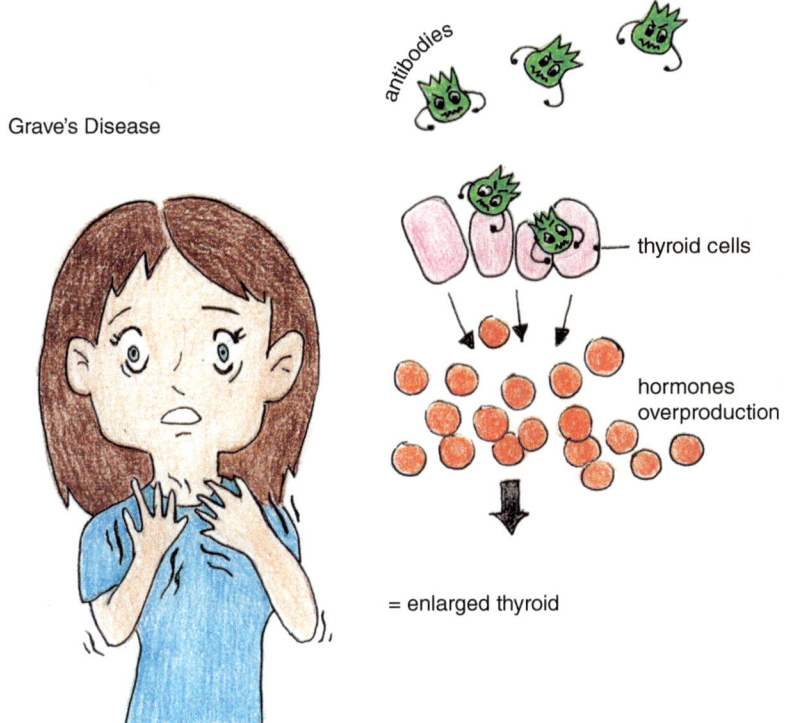

Grave's Disease

antibodies

thyroid cells

hormones overproduction

= enlarged thyroid

In this disease, the immune system recognizes the thyroid as an outsider and attacks it, producing the *TRAb antibodies* located on the thyroid cells. These cells, stimulated by the TRAb antibodies, are induced to increase the production of thyroid hormones, resulting in those symptoms of hyperthyroidism listed before (insomnia, anxiety, irritability, accelerated heartbeat, warm skin, sweaty hands, and so on). Moreover, the thyroid follicles, since continuously stimulated to produce hormones, start to enlarge and could form a goiter.

But it doesn't end here! In this autoimmune disease, antibodies can also attack other organs besides the thyroid, for instance, the eye orbits. Quite often, swollen and protruding eyes are the most evident sign of this condition. The wide-eyed effect is due to the fact that the upper eyelids tend to retract, leaving the white part around the pupil exposed. In addition, there can be sight problems, such as double or blurred vision, accompanied by a burning sensation or pain inside the eye. Not only does it become an aesthetic problem, but if neglected over time, it can severely compromise our vision!

The endocrinologist will prescribe a treatment that blocks the excessive production of thyroid hormones, to get out of the state of hyperthyroidism with all its unpleasant consequences.

The treatment may consist of anti-thyroid drugs, as well as a surgical intervention, or a radioactive iodine-based therapy.

## 2.4    Nodules

We very often find nodules that have emerged on the thyroid: a part of its tissue, sometimes even very small, enlarges to create a ball, which the larger it becomes the more possibilities it can be bothersome for us as we swallow, speak, or breathe. The presence of nodules should not scare us; it is a very common circumstance and it usually doesn't interfere with the gland activity. Remember that they always need to be kept under control!

Healthy thyroid                                      Thyroid nodule

## How to Find Them?

Nodules are not easy to find and in many instances, they are discovered by chance, since symptoms may not even appear, or rather because they are so small that they escape both sight and touch. We need a good and experienced "investigator" to find them. Luckily, we have our endocrinologist who knows well how to move! First things first, the doctor palpates our neck during the physical examination, keeping in mind that only 7–15% of nodules are large enough to be found in this way.

Thenceforth, we avail ourselves of an examination called **Ultrasound** (US), which ideally manages to find even the tiny ones!

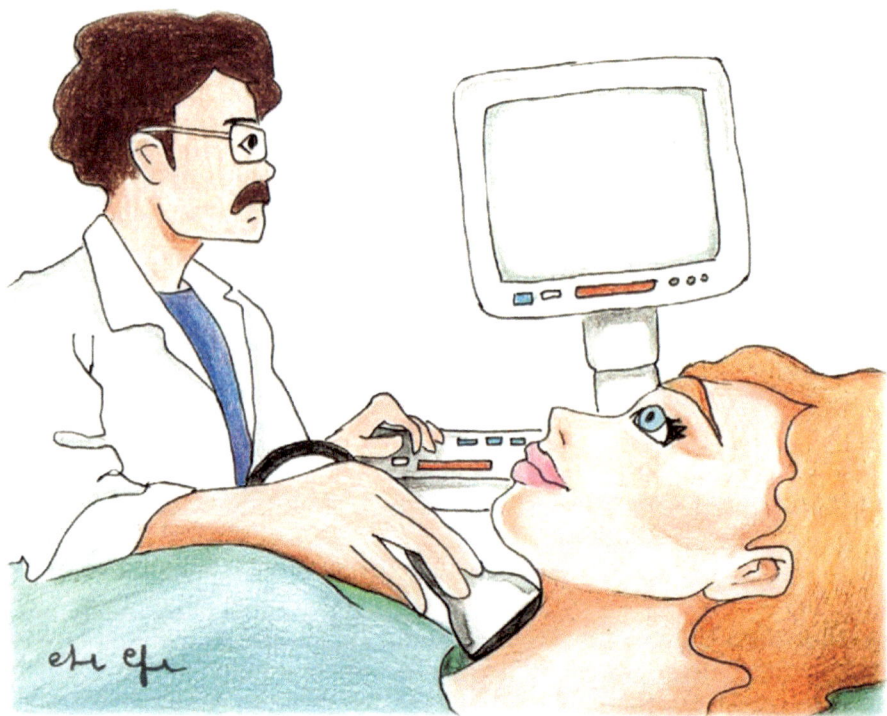

## What to Do?

Once identified, we need to decide how to manage them. First of all, we need to find out "which side they are on," whether on the "good side" (**benign nodules**) or the "bad side" (**malignant nodules** or **cancerous**).

In order to understand their nature, the endocrinologist might ask us to run other exams, such as thyroid hormones, calcitonin, thyroid scintigraphy, or a fine needle aspiration. In the case of benign nodules, usually, there is no need to do anything, just check them from time to time.

In the second case, however, we need to promptly intervene to remove them, before they can do damage! On the bright side, the "bad" ones are less common (only 5–7%). To remove them, it might be necessary to remove part or even all of the thyroid gland. And how can we do without it? Fear not, its activities will be performed by levothyroxine, a synthetic thyroid hormone that has the same functions as those naturally produced by the thyroid.

**Remember That…**

√ Like all our organs, the thyroid can also become ill. Having a nodule, or more than one, is a fairly common circumstance and affects approximately 60–70% of the population, with 60 thousand new cases diagnosed each year in Italy. Nodules are almost always benign, and in most cases, they are not serious and do not cause any disturbances. Only 5–7% of them exhibit characteristics of malignancy. It is important to note that a thyroid nodule is not necessarily a tumor!

√ Another thyroid disease is goiter, which presents itself with an increase in the size of our gland. There are two main types of goiter: diffuse goiter, where the entire thyroid is enlarged, and its surface is smooth to the touch, and nodular goiter, where the size increase affects a part of our thyroid, and nodules are present, making the surface irregular to the touch. Goiter is usually non-toxic and allows the thyroid to function regularly, but it should be monitored because over time it could transform into toxic multinodular goiter. However, the two main manifestations of thyroid diseases are hyperthyroidism and hypothyroidism.

√ When our gland works more than necessary and produces too many hormones, it is referred to as hyperthyroidism, a condition that causes a series of symptoms, including nervousness, anxiety, hyperactivity, weight loss, and rapid or irregular heartbeat. Among its possible causes is Graves' disease, an autoimmune disease that can occur at any age and presents with swollen and protruding eyes. Like all autoimmune diseases, this condition produces so-called autoantibodies, which are antibodies that fight against the normal components of our body, recognizing them as an external threat.

✔ When our thyroid becomes sluggish and does not produce enough thyroid hormones, it is called hypothyroidism: fatigue, excessive sensitivity to cold, dry skin, and fragile hair are evidence of it. Anyone can suffer from hypothyroidism, although it is more common among women, especially over the age of 60. Among its possible causes is Hashimoto's thyroiditis, an autoimmune disease that is the leading cause of thyroid dysfunction in areas with sufficient iodine intake. It can be present in up to 15% of women and up to 5% of men.

✔ Among thyroiditis, finally, we have subacute thyroiditis (De Quervain's thyroiditis) and acute thyroiditis. The first occurs when the thyroid infection is caused by a virus and manifests with pain in the front of the neck, sometimes with fever and physical fatigue. The second, instead, is caused by a bacterial infection: it presents the same symptoms as subacute thyroiditis and, in addition, can manifest with warm and reddened skin on the neck.

✔ For all these thyroid diseases, there is a treatment: the endocrinologist will be able to advise on the most suitable one for us and our clinical situation!

# Thyroid at All Ages

**3**

**Thyroid at All Ages**

Growth is one of the most important functions performed by the thyroid: it is the one responsible for all those mechanisms that allow our body to develop harmoniously, both from a physical and mental point of view.

Therefore, it is extremely important to "grow up" in the best way as our gland keeps us company throughout our life, in its different phases, safeguarding us from diseases and dysfunctions. Let's see everything in detail!

D. Pace, *Thyroid: The Butterfly of Metabolism*,
https://doi.org/10.1007/978-3-031-55276-2_3

## 3.1    When the Thyroid Is Young

In newborns and children the production of thyroid hormones is molding for the first phases of growth of our organism.

When we enter puberty, they become fundamental, since they play an important role in the complex changes of this transition phase towards adulthood (characterized by the production of many hormones and transformation of organs and tissues), contributing to complete our psycho-physical growth.

### "I" for Iodine and Stands for Intellect!

Once again, iodine is there to support our thyroid in this incessant work (as we have seen is an indispensable element). It has been ascertained that iodine deficiency in the early stages of life leads to a diminished functionality of the thyroid, which is associated with levels of intelligence quotient (IQ) much lower than the average.

In some areas of the world, however, where iodine deficiency is pronounced, it is essential to make use of specific supplements.

### What If the Thyroid Gets Sick?

Less frequently than adults, children and teenagers thyroid can suffer from the same disorders. Both **hypothyroidism** and **hyperthyroidism** symptoms may appear at any age. In particular, hypothyroidism is the most common endocrine disease in children and adolescents. It affects girls more than boys and is usually the result of autoimmune thyroiditis or iodine deficiency. In other cases, it may appear at birth or in early childhood. Treatments are the same as per adults, obviously adapted to each age group. Hyperthyroidism, on the other hand, is largely due to Graves' disease. There are also cases where a *hyper-active* thyroid is the result of many teenagers' bad habits, who—worried about their physical appearance— resort to supplements or drugs to lose weight, such as seaweed that comes with a naturally high iodine content or even medicines used for the hypothyroidism treatment, such as levothyroxine; it goes without saying that, in these cases, there's no real need or, even worse, it's outside medical control. Being young is not a good excuse for not using your head!

On the contrary, when you are a step away from adulthood and you can take full responsibility for your actions, you must show matureness by taking drugs only if necessary (in this particular case, following the indications of our endocrinologist) and staying away from do-it-yourself remedies. Health is not a joke!

## 3.2    When the Thyroid Ages

As we have seen, the thyroid is essential for our growth as newborns, children, and teenagers, but it is equally fundamental in adulthood as it continues to perform important functions for our body. For instance, it regulates metabolism, heart rate, body temperature, and brain abilities (i.e., memory and attention) and even influences nails and hair growth. The thyroid accompanies us in every stage of life, as it grows and ages with us!

Thyroid ages with us

### The Grandparents'Thyroid

As we age, we must continue to take care of our gland; this is especially true, due to the fact that, when you are over 65 years old, everything becomes kind of blurred, the signals that our body sends us are weaker and can easily be confused, or the symptoms can overlap with other complaints.

Some may even argue that aging itself is a consequence of a tired thyroid, which after working non-stop for a lifetime, decides to retire. It's not quite right, but certainly, over time, there are many changes, and as a result, we could feel we are no longer as fresh as a daisy, as we once were.

The thyroid is not immune from aging, still, it does not cease its work, on the contrary, it tries to adapt to changes and to compensate, if possible, to always keep everything in balance.

## Hypothyroidism

Let's now see how hypothyroidism and hyperthyroidism, the two thyroid conditions we've already come to know, can manifest in old age.

**Hypothyroidism** often does not come to light with the classic symptom of weight gain, we might notice weight loss instead, as a consequence of a loss of appetite.

Other typical warning signs, such as fatigue, dry skin, hearing loss, less mental clarity with some small memory gaps, and low mood, can be mistaken for characteristics of people who are getting older.

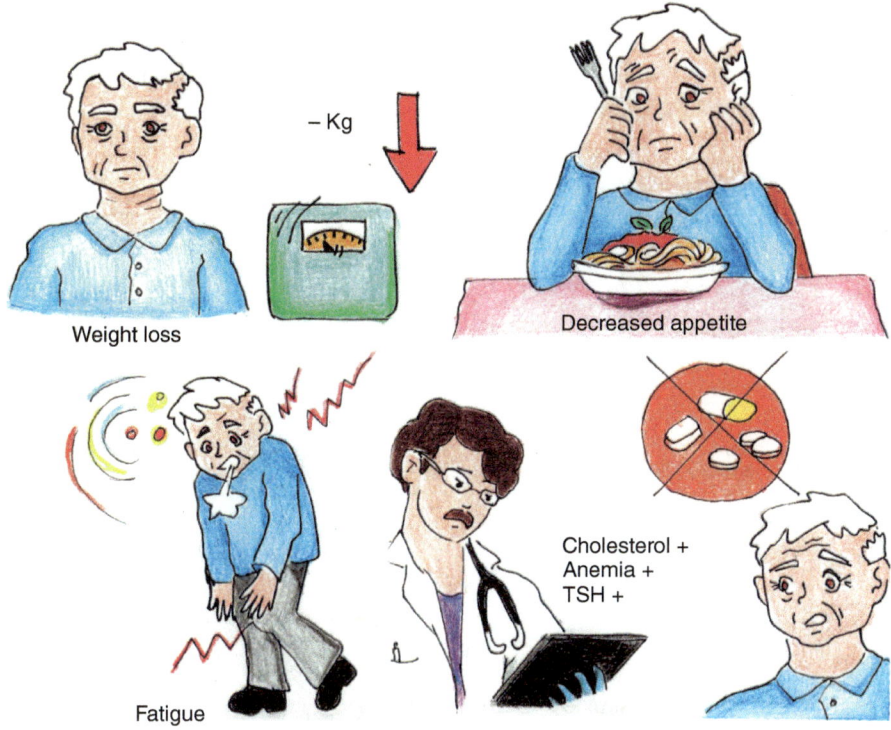

Weight loss

Decreased appetite

Fatigue

Cholesterol +
Anemia +
TSH +

None of these symptoms are particularly suspicious per se until we run some blood tests and anemia, or high cholesterol levels, come into play; these could be a sign of hypothyroidism.

In milder cases as well as in older people, the endocrinologist might even decide to not prescribe any medicine that could cause more side effects than benefits. It is thought that hypothyroidism in elder people has no harmful consequences; on the contrary, it exerts a protective effect that prolongs life! It's as if the thyroid decreases its work a little, enough to save energy so that it can continue to work longer.

## Hyperthyroidism, the Subtle Enemy

If not recognized at an early stage, **hyperthyroidism** can lead to various complications.

The diagnosis, however, may be often difficult, since the typical signs of this dysfunction are not so evident. Often, there might only be an accentuation of a symptom, such as weight loss, brain fog, or cardiovascular problems. Even in this case, these warning signs could be attributed to other diseases as well as the normal aging process. Sometimes, it could be the manifestation of untreated hyperthyroidism, causing symptoms such as arrhythmias or hypertension that reveals its presence. With age, moreover, bones tend to shrink in size and density making them prone to fracture, thinking skills, mood swings, and behavioral disorders.

The physician will recommend the most appropriate treatment, taking into account the patient's age, medical history, and the degree of hyperthyroidism: drugs are usually prescribed and in exceptional cases only, surgery may be recommended.

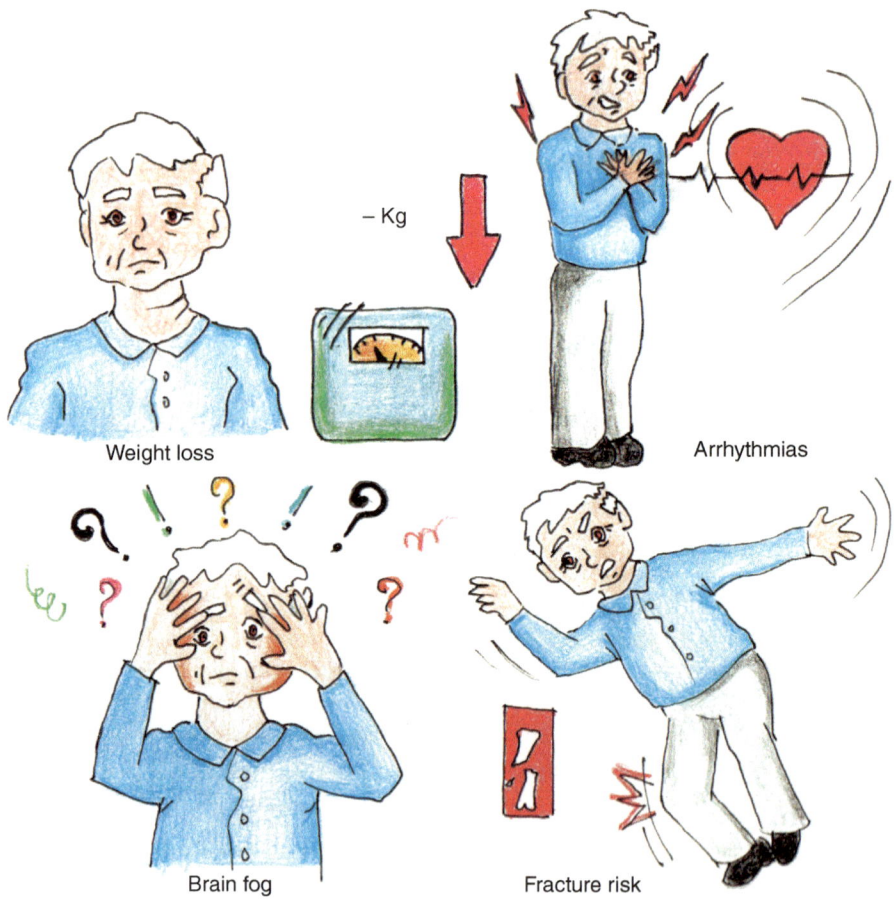

Weight loss

Arrhythmias

Brain fog

Fracture risk

## 3.3    When Mum Is Expecting

In pregnancy, the thyroid will be called to perform a supertask! Throughout the pregnancy, the thyroid will slightly enlarge and it will be called for "extra" production of hormones due to all the changes in the woman's body.

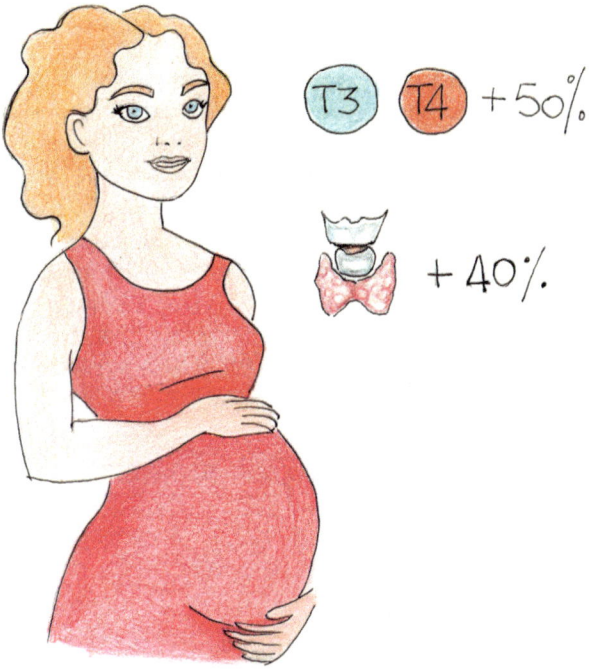

## What Happens During Pregnancy

Indeed, thyroid hormones play a crucial role in normal development of the baby's brain and nervous system. The woman's body changes and the baby bump will gradually grow, while inside the baby is developing. The thyroid size can increase by up to 40% during pregnancy, while simultaneously increasing the production of thyroid hormones by 50% and, in parallel, the need for iodine.

But that's not all. In case, there are dysfunctions from which the mum has suffered in the past or that, until that moment, she had not noticed because the symptoms were not so obvious (in this case we speak of **latent** infection), these could (re) appear any moment, affecting the progression of pregnancy, as well as the fetal development, and the postpartum period. Therefore, the mother needs to take particular care of her thyroid, as the repercussions can be felt not only on her, but on the baby too since the two of them are bonded to each other!

## Mum, Remember the Iodine!

Taking the right amount of iodine is particularly important during pregnancy. Since the baby she is carrying in her belly is not able to produce thyroid hormones yet, mum will have to do it for the little one too, and that's why she will need more iodine "to manufacture" them. In particular, this happens in the first 3–4 months, when she will have to take 50% more than before; if this is not enough, she can always rely on the use of iodized salt or supplements. Iodine is crucial, especially to ensure the correct development of the baby's nervous system. This micronutrient is also important after birth because the newborn will need it for proper growth. If the mother breastfeeds, she will transmit it to the child through her milk, which will be as rich in iodine as she takes herself. This implies that she will have to continue to take it in greater quantities, exactly as she did during pregnancy.

If the mother resorts to formula milk, she can use specific infant formulas containing the right doses of iodine. Then, around the first year of age, when the toddler will gradually start eating other foods in addition to milk (this phase is called **weaning**), he will get iodine both from food and from iodized salt, just like you do!

## A Great Upheaval

Pregnancy is a particularly beautiful moment for every woman, almost "magical" we could say, but it also involves great hormonal changes! One of these two things can happen: if the mother suffered from autoimmune diseases before pregnancy, these can even improve, and in some cases—for example, when one suffers from Graves' disease—the doctor may advise her to suspend treatment; or, even if she has never suffered from them in the past, disorders such as hypothyroidism or hyperthyroidism may appear, right when she is expecting. But don't worry! The good news is that treatments planned for these disorders can also be taken during the gestation period.

## Too Many Hormones in Pregnancy

Some signs and symptoms of **hyperthyroidism** (too many thyroid hormones) may occur during pregnancy, including heat intolerance and fast heartbeat; other signs are more specific, including weight loss, goiter, and the emphasis of some of the "more classic" disturbances, (e.g., excessive sweating and fatigue).

Whether the patient's medical history or the mother has already been diagnosed with hyperthyroidism in the past, it becomes necessary to check the functionality of the thyroid. The consequences could be serious for both mother and child. If controlled, hyperthyroidism won't cause damage. In most cases, the doctor may prescribe antithyroid medicines, which block the production of the hormones. They can also be taken during pregnancy to calm the hyperactivity of the thyroid and prevent too much of the mom's thyroid hormone from getting into the baby's bloodstream. This goal needs to be achieved as soon as possible, to maintain the right levels of hormones with the minimum dosage.

For this reason, the mother needs to run frequent blood tests (at least once a month) throughout the gestation period, checking the weight, heart rate, thyroid size, and the number of hormones in the blood; in this way, the situation will always be under control and the physician will be able to adjust the therapy accordingly. Only in exceptional cases, surgery may be needed to remove the gland or part of it. Radioactive iodine treatment is not an option for pregnant women. If the mother suffered from hyperthyroidism in the past and therefore followed the treatment provided in these cases, she will probably have to change the dose or even stop taking it, since hyperthyroidism caused by Graves' disease, as we have seen, improves in pregnancy.

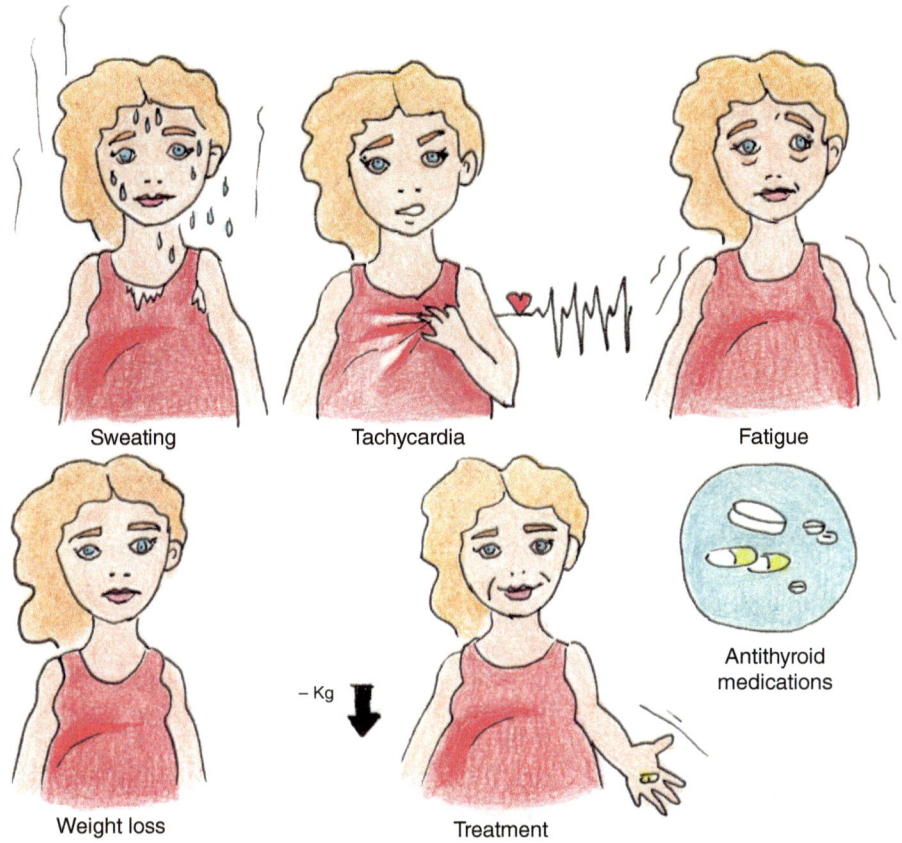

Sweating            Tachycardia            Fatigue

Antithyroid
medications

Weight loss          Treatment

## When Hormones Are Too Little

**Hypothyroidism** is characterized by a low production of thyroid hormones. Because thyroid hormones are very important to the baby's brain and nervous system development, especially during the first trimester, untreated hypothyroidism can lead to complications for the mother, including hypertension or the risk of preterm birth. Diagnosing hypothyroidism during pregnancy is often proved to be a difficult task, since the signs, such as cold intolerance, hair loss, brain fog, and fatigue, are also common symptoms in pregnancy. What causes hypothyroidism in pregnancy? Hashimoto's thyroiditis. The doctor may do some blood tests, to look for the presence of certain antibodies. Other causes could be the medical treatment the mother followed before pregnancy to treat hyperthyroidism, or the surgical removal—partial or total—of the gland. The physician will certainly review the symptoms and do some blood tests to measure the TSH levels, especially if the woman's medical history suggests thyroid diseases as well as autoimmune diseases, such as celiac disease, rheumatoid arthritis, or vitiligo in the family. It would be appropriate to do this even before *planning* a pregnancy! If the mother discovers that she suffers from hypothyroidism **during** pregnancy, the endocrinologist will most likely prescribe a synthetic thyroid hormone (levothyroxine) treatment, which is obtained in the laboratory, and he/she will monitor thyroid function throughout the pregnancy. If the mother had hypothyroidism **before** getting pregnant, however, she will probably need to increase the dosage. She may continue the treatment even during breastfeeding.

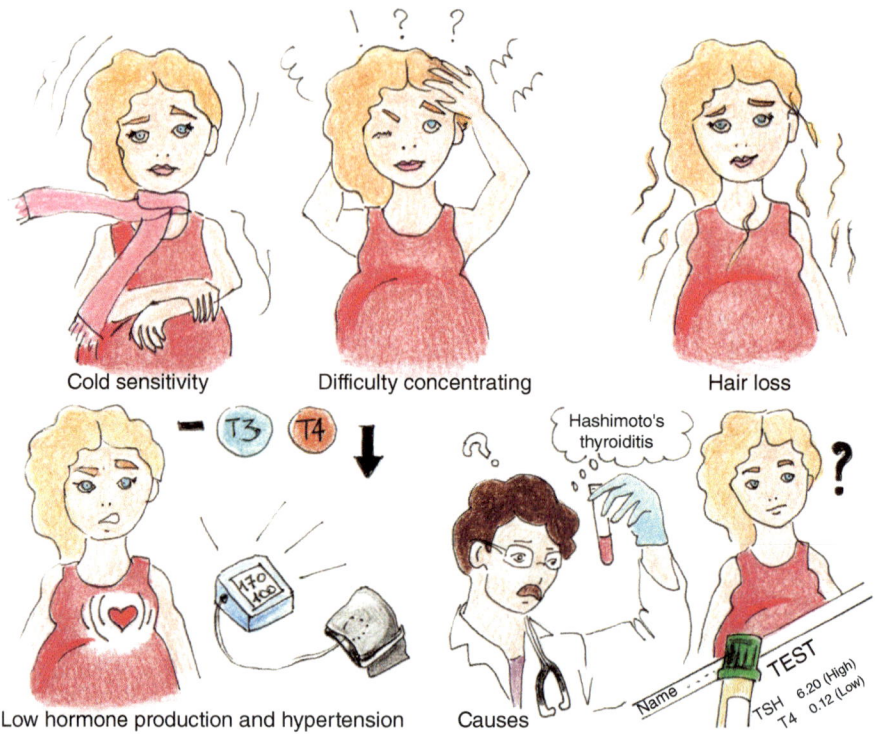

Cold sensitivity                Difficulty concentrating                Hair loss

Low hormone production and hypertension        Causes

## Hypothyroidism and Miscarriage

Uncontrolled hypothyroidism as well as autoimmune thyroiditis is considered among the most frequent causes of miscarriages. Scientific studies have shown that women with hypothyroidism who follow a hormonal replacement therapy—a treatment that helps them to fill the thyroid hormone deficiency—and who have a TSH range above 2.5 mU/L at the beginning of pregnancy, are exposed to a higher risk of spontaneous abortion.

Other researchers have highlighted a link between the latter and antithyroid antibodies: patients positive for such antibodies have a double risk of spontaneous abortion than those who are negative.

That's why it is necessary to seek a specialist and consult your endocrinologist as soon as you decide to have a child; sometimes one realizes to have a thyroid problem only after having an abortion.

## The Discovery of a Thyroid Nodule

What if a woman discovers the presence of a thyroid nodule right when she is expecting? It could certainly cause some apprehension (it's normal!) but there is no reason to panic. Only a small minority of nodules (less than 7–15%) are malignant and pregnancy does not affect their evolution.

Among the investigations to understand the nature of a nodule, Scintigraphy is the only one to be avoided when you are pregnant.

In most cases, there is no need to intervene, and if surgery is necessary, the risks are minimal, both for the mother and the child. The discovery of a thyroid nodule during pregnancy, therefore, should not worry excessively, but it is still necessary to keep the situation under control.

## After Delivery

It can happen, after giving birth to the baby, that the new mother finds herself facing a difficult period, in which she feels a bit down.

It's her body's reaction to the great upheaval of hormones and it happens to about 80% of women who gave birth. These are mild symptoms of depression that usually resolve spontaneously in a couple of weeks. Sometimes, however, this state of sadness can last longer and the symptoms can be more pronounced (it happens to 10 mums out of 100!).

Evidence in the literature has suggested a correlation between **postpartum depression** and thyroid antibodies. As a matter of fact, in most cases, thyroid dysfunctions, such as hypothyroidism and, although less frequently, hyperthyroidism, exhibit comorbidity with various psychiatric disorders, including mood disorders and maternal depression. That's why we need a "double treatment": one for thyroid disorders and the other one will be for the blues.

It's very important to constantly monitor the overall mum's health and wellness, in pregnancy and after giving birth, in order to be ready to intervene as fast as possible, should the need arise, and restore balance.

**Remember That...**

√ The thyroid accompanies us throughout our life's journey. When we are infants and children, thyroid hormones are crucial for the early stages of our body's growth and remain equally important during adolescence. They play a crucial role in the complex changes that occur in the body during this period, contributing to the completion of regular psycho-physical development.

√ Although less frequently, children and adolescents' thyroid can experience the same disorders as adults. Hypothyroidism is the most common disease and is usually a consequence of autoimmune thyroiditis or iodine deficiency. Conversely, hyperthyroidism is largely due to Graves' disease. Among its causes, incorrect habits of many teenagers resorting to supplements and weight loss medications should not be excluded. For all types of thyroid inflammation and dysfunction, the treatments are the same as for adults but adapted to their age.

√ As we age, the attention we dedicate to our gland should not diminish. This is particularly important because, in old age, we tend to confuse thyroid disease symptoms with normal age-related disorders. Hypothyroidism, for example, contrary to what happens in other stages of life, presents with weight loss, a direct consequence of decreased appetite. Along with it, there may be fatigue, hearing loss, reduced mental clarity, etc. It is good to know that in milder cases of hypothyroidism, the endocrinologist may decide not to administer medication, as it may have more side effects than benefits.

√ Hyperthyroidism, on the other hand, is more hostile and difficult to diagnose because the typical signs of this dysfunction do not always appear. Sometimes, it is the manifestation of a consequence of untreated hyperthyroidism, such as arrhythmias or hypertension that reveals its existence. Age, the presence of other diseases, and the degree of hyperthyroidism will guide the doctor in choosing the appropriate treatment: usually, medication is used, and surgical interventions are only considered in exceptional cases.

√ Thyroid hormones play a crucial role during pregnancy, a period that requires extra production (50% more!) to support the changes in the mother's body and the growth of the child. Taking the right amount of iodine is absolutely necessary, even in the months immediately after childbirth. This can be done through the use of iodized salt or, if that is not enough, through specific supplements.

√ Pregnancy is a beautiful moment but brings with it significant hormonal upheaval that could result in two different situations: if the mother had autoimmune diseases before pregnancy, these could even improve. In some cases, such as when suffering from Graves' disease, the doctor may advise discontinuing the therapy. Alternatively, even without having suffered from them in the past, disorders such as hypothyroidism or hyperthyroidism may appear.

√ Hyperthyroidism manifests with symptoms that may be common to those of pregnancy, such as intolerance to heat and a rapid heartbeat; others are more specific, such as weight loss, goiter, and accentuation of "classic" momentary disturbances, such as excessive sweating and fatigue. If controlled, this dysfunction does not cause harm and can be treated with antithyroid medications even during pregnancy. If the mother already had hyperthyroidism before pregnancy, she may need to change the dosage of the medication she is taking or even suspend its intake!

√ Hypothyroidism during pregnancy has more critical aspects for both the mother and the child, and its diagnosis is often quite challenging as it manifests with symptoms similar to those of the period (intolerance to cold, hair loss, difficulty concentrating, and fatigue).

√ Among the causes of hypothyroidism during pregnancy are Hashimoto's thyroiditis, the pharmacological therapy that the mother followed before pregnancy to treat hyperthyroidism, or the surgical removal of a nodule. There is also a cure for this dysfunction: if the mother discovers she has hypothyroidism during pregnancy, the solution is treatment with thyroid hormone (L-thyroxine). If the mother already had hypothyroidism before pregnancy, she can continue her therapy calmly, although she may need to increase the doses. Diagnosis, analysis, and therapy will be prescribed and elaborated by our endocrinologist! And if during pregnancy, the mother discovers she has a thyroid nodule?

# What to Do If the Thyroid Gets Sick

# 4

**What to Do If the Thyroid Gets Sick**

The endocrinologist, the medical doctor who checks the overall thyroid health, can use various tools for his *investigations* as they can help him understand what is happening in our body. They have become effective weapons, some of them simpler than others, thanks to technology, which keeps evolving. Are you ready to know them?

© The Author(s), under exclusive license to Springer Nature Switzerland AG 2024
D. Pace, *Thyroid: The Butterfly of Metabolism*,
https://doi.org/10.1007/978-3-031-55276-2_4

## 4.1    Blood Tests

A simple blood test can tell us a lot about our butterfly health. How is it done?

The nurse takes a few drops of our blood with a very thin needle from a blood vessel in your arm, usually from the inside of the elbow or the back of your wrists, where the veins are close to the surface; the needle is connected to a syringe or a blood sample bottle.

The sample will be sent to the laboratory for analysis.

To understand if your thyroid is functioning well or if maybe it is working abnormally, too much or too little, the laboratory technicians and analysts will evaluate the amount of TSH from the blood sample. If TSH levels are high, we could have hypothyroidism, if too low they could indicate hyperthyroidism... but we have already learned it in the previous pages!

Along with TSH, the amount of FT4 present is also evaluated providing indications on the amount of thyroid hormones.

Blood tests
results
TSH ........
FT4 ........

Actually, **hypothyroidism** could be diagnosed by measuring the TSH, since it begins to increase even when thyroid hormones are still normal; this is the case of *subclinical hypothyroidism*, which is the initial form of this dysfunction.

In the case of *frank hypothyroidism*, TSH levels are elevated, whereas fT3 and fT4 levels will be lower than average: this is the most advanced condition of hypothyroidism.

Another value that can be taken into account is the presence of auto antibodies, which could indicate that an autoimmune disease is the cause of hypothyroidism.

In particular, to determine if **Hashimoto's thyroiditis** is the cause of hypothyroidism, your healthcare provider will look for two types of antibodies called anti-thyroid peroxidase (TPO) and anti-thyroglobulin (TG). In Hashimoto's thyroiditis, the presence of the first type is in 90% of cases, while the second one is in 60%.

If, on the other hand, **hyperthyroidism** is suspected, the evaluation will be on the FT3 and FT4 levels in the blood.

TSH is reliable data for hyperthyroidism too, since its level decreases immediately, even when the thyroid hormones levels start slightly increasing. This is precisely why we do not perceive its severity -if hyperthyroidism is in an initial phase or if it is already in an advanced form- until we see what the situation of thyroid hormones is.

Low TSH levels and high levels of FT3 and FT4 usually indicate an overactive thyroid.

However, this disorder is only a symptom, and the endocrinologist will start investigating what has pushed the thyroid to produce all these excess hormones, prescribing an ultrasound or a scan.

Blood tests may be useful to check if the parathyroids are healthy since they perform an important job in collaboration with the thyroid. In this case, the levels of calcium, PTH, and vitamin D to understand if the parathyroids "are well" and if the calcium metabolism is functioning properly.

## 4.2   Thyroid Ultrasound

Ultrasound is a valid tool that allows the endocrinologist to see what is happening under our skin, in the internal organs, thanks to a probe that slides on the surface of the body and, as for the thyroid, along the neck.

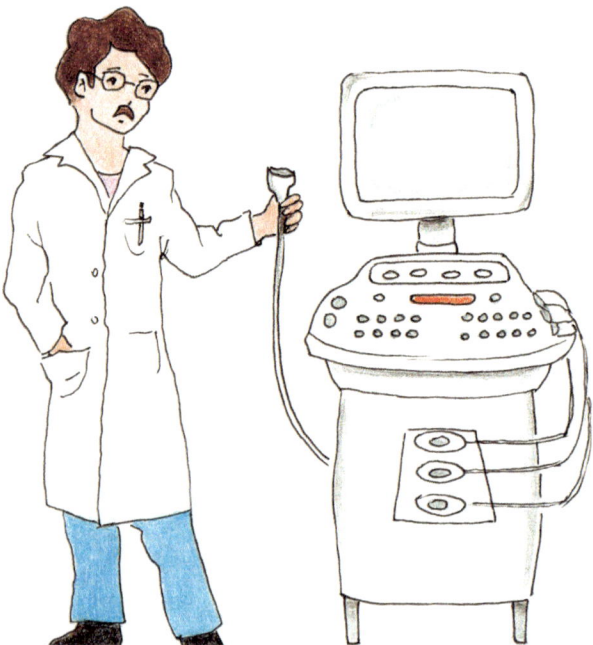

The probe emits ultrasound waves that our ears are not able to hear; however, they are perfectly perceived by our internal tissues. The echoes bounced off the thyroid are converted by a computer program into a black and white image.

To tell the truth, for those who are inexperienced it takes a bit of imagination to glimpse "the wings of our butterfly" in the black and white images of the ultrasound, but a good ultrasound doctor will be perfectly able to interpret them. He will evaluate the shape, morphology, and volume of the gland, identifying inflammation or the presence of any nodules with extreme precision, managing to find even the tiny ones, whose diameter is 2–3 mm.

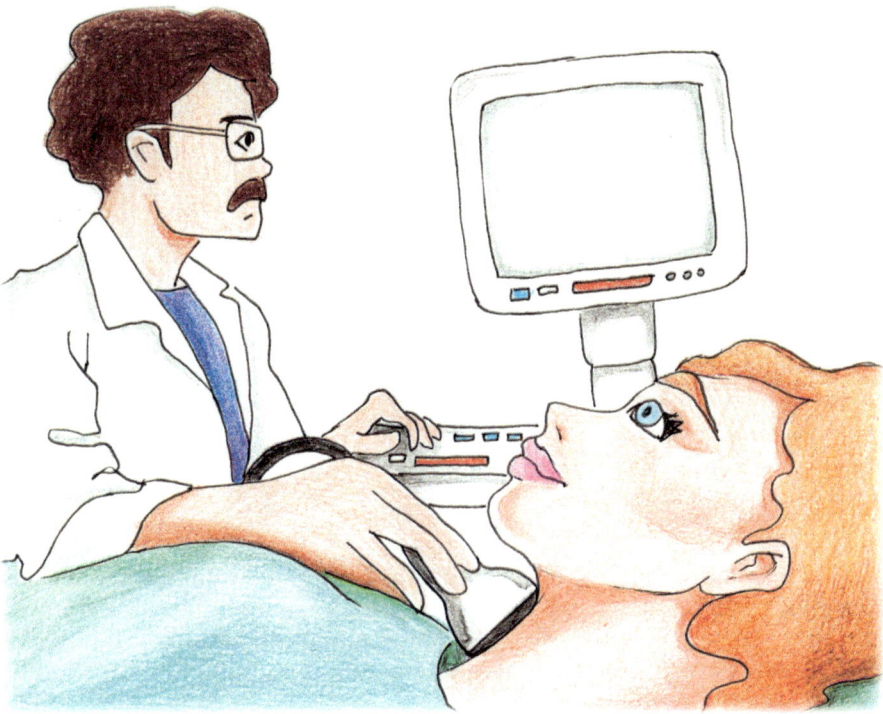

At the same time, he will be able to see if they are benign or if they fall into that small percentage of malignant nodules. In this case, to better understand their nature, he will ask to delve deeper with a further examination: the ultrasound-guided fine-needle aspiration. Let's see together what it is!

## 4.3   Ultrasound-Guided Fine-Needle Aspiration (FNA)

Fine-needle aspiration is a procedure, carried out by the endocrinologist who takes a sample of the tissue in the nodule and sends it to the cytologist laboratory to evaluate their nature: benign or malignant.

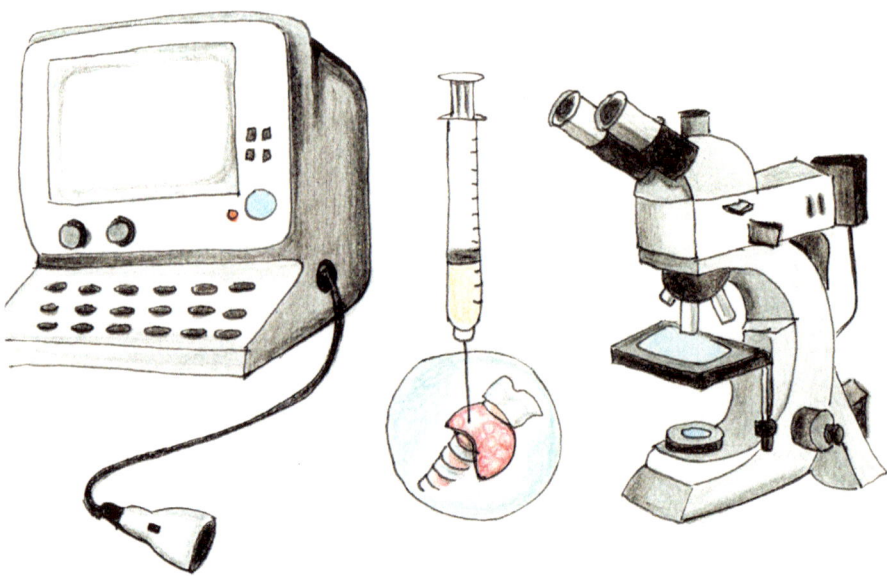

To perform this examination, the doctor will ask us to lie down on an examination couch with our head resting on a pillow, so we can recline the neck backward. A very small needle will be inserted through the exact point where the nodule is located with the assistance of ultrasound images and it will be "jiggled" back and forth to take the sample.

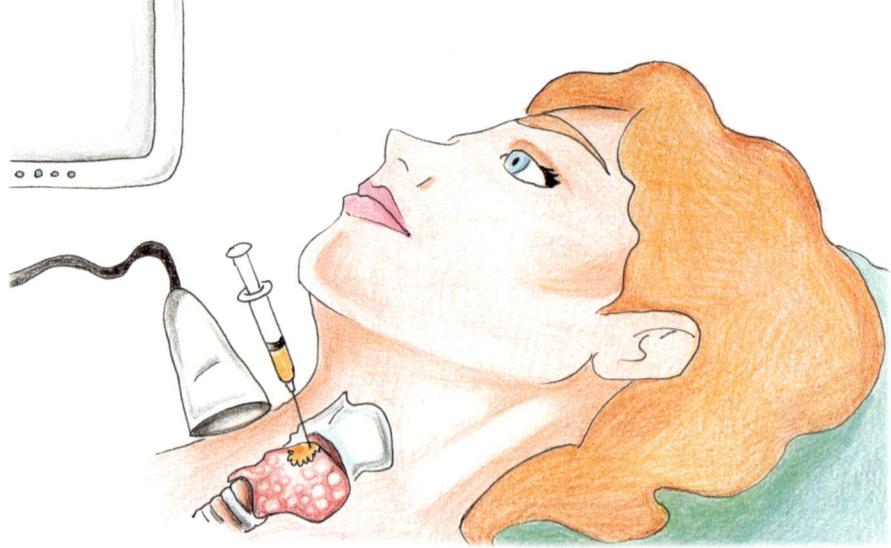

The puncture may cause a bit of discomfort, but it doesn't last long: 30 seconds tops!

Sometimes the neck can remain sore for a while and in the following days we may see the formation of a small bruise. It's normal, it will fade very soon! In the meantime, the cells taken will already be brought to the laboratory for analysis, allowing us to understand the nature of the nodule, whether it falls into the "good" or the "bad" category.

## 4.4 Thyroid Scintigraphy

Finally, thyroid scintigraphy (or scan) was a procedure widely used in the past, while today we rely on it only for specific cases. It is mainly used to determine the cause of hyperthyroidism after blood tests have revealed a low level of TSH. Therefore, it serves to confirm the diagnosis of Graves' disease.

It is also used when the ultrasound has detected a nodule; if that's the case, the scan allows one to distinguish a *hot* nodule, which indicates hyperfunction, from a *cold* one, indicating hypofunction.

Thyroid scintigraphy is administrated, through an injection, or swallowing a small dose of a particular radioactive medicine, called **radioisotope** (*technetium* or *iodine 123 or 131*): this directly communicates with our thyroid, prompting it to emit response signals, the so-called *uptake* (this test is also called RAIU, **radioactive iodine uptake**).

After taking the radiotracer, the doctor will have us lie down on the exam table with our heads tipped backward and necks extended; he will pass a special scanner near the neck capturing the radiation emitted by the thyroid.

The results, printed on an X-ray film, reproduce a real map of our thyroid:

- If the gland is perfectly functioning, its unique butterfly-shaped image will be reproduced, with uniform density and color, as it emits homogeneous and diffuse signals along its entire surface;
- A thyroid nodule that functions too much, instead, appears as a very dense and colored area compared to the rest of the thyroid. We'll have then a *hot nodule*;
- A less functioning nodule, on the other hand, does not respond to the radiopharmaceutical uptake, therefore it appears as a lighter and less colored area. In this case, we'll have a *cold nodule*.

Over 90% of hot nodules are benign; therefore, it is not necessary to perform a fine-needle aspiration to verify their nature. The cold nodules, on the other hand, have a suspicion of malignancy (SFM) in 8–25% of cases, which is why it is recommended to proceed with the fine needle aspiration procedure.

All this information provided by the scintigraphy is particularly useful in determining how hyper- (in excess) or hypo- (below/under) functioning the nodules are, and everything will be evaluated by the endocrinologist for a complete diagnosis, along with ultrasound, needle aspiration, and blood test results. The teamwork between all these investigations will allow us to have a better picture, allowing our "detective" to assess the thyroid clinical situation in the best way possible.

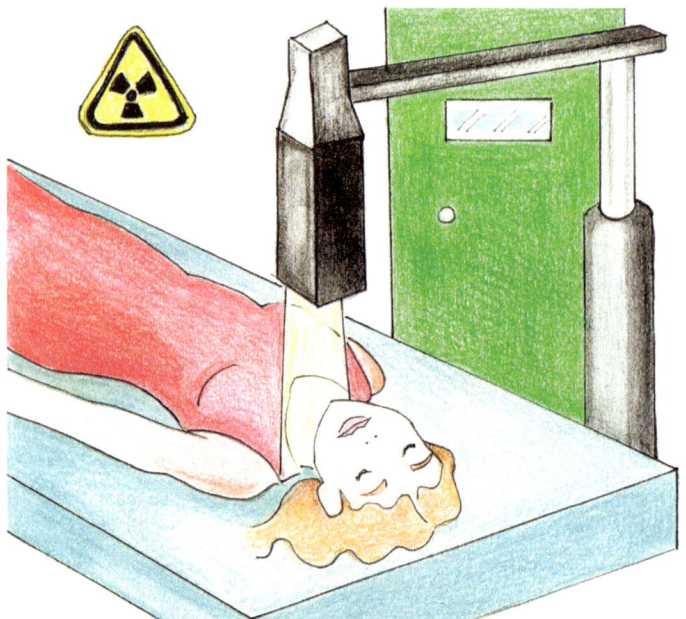

**Remember That…**

✓ To understand if our thyroid is healthy and functioning well, the endocrinologist can rely on various investigative methods; some are simpler, others more complex.

✓ Usually, a simple blood test that evaluates TSH and thyroid hormones can tell us a lot about the health of our gland: if TSH levels are high, we are in a hypothyroidism situation; if too low, it indicates hyperthyroidism. By also checking FT4, it is evaluated whether the thyroid is producing too many or too few hormones.

✓ Other tests that may be considered include the presence of autoantibodies, which indicate that the cause of hypothyroidism is an autoimmune disease. In the case of Hashimoto's thyroiditis, anti-TPO and anti-Tg antibodies will be assessed.

✓ In the case of hyperthyroidism, the anti-TSH receptor antibodies (TRAb) should also be evaluated to understand if the cause is an autoimmune disease called Graves' disease.

✓ Blood tests can also be useful to check the health of the parathyroids: in this case, the quantities of calcium, PTH, and vitamin D are analyzed to understand if calcium metabolism is functioning properly.

✓ Among the investigative tools is also ultrasound, which allows seeing what is happening in our thyroid through a simple probe that moves on the surface of the neck and emits ultrasound. These, through processing by a computer program, provide images of our internal organs and tissues, through which the ultrasound doctor can assess the shape, morphology, and volume of the gland, identifying inflammation or the presence of any nodules.

✓ To understand the nature of thyroid nodules (benign or malignant), ultrasound-guided fine-needle aspiration is used. The specialist will have us lie down with our neck tilted backward, and through the ultrasound probe, will locate the exact point where the nodule is and insert a thin needle through which he will collect cells for analysis under a microscope. The puncture may be a bit uncomfortable, but it will only take a few seconds!

✓ Thyroid scintigraphy is also among the investigations that the endocrinologist can request to assess the state of our butterfly. It is used after blood tests have shown a low amount of TSH, and ultrasound has detected the presence of a nodule.

✓ It is performed by administering a small dose of a radioisotope, a specific radioactive drug capable of "communicating" with our thyroid. The results of the scintigraphy, printed on a radio-graphic film, reproduce a real map of our thyroid.

# How and When to Treat the Thyroid

**5**

**How and When to Treat the Thyroid**

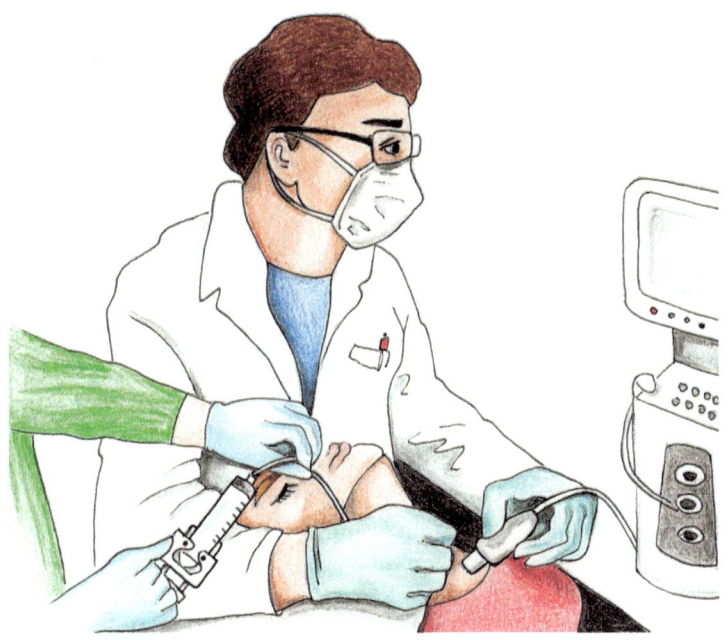

As we've already seen in the previous chapter, the endocrinologist will examine all the data. Following the diagnosis, which means that, he will clarify our butterfly and endocrine condition, suggesting a treatment plan. Let's see all the possible solutions!

## 5.1    Thyroid Medications

Underactive thyroid (hypothyroidism) is usually treated by taking **Levothyroxine** or **L-thyroxine**, often abbreviated to **LT4,** which is a synthetic version of the T4 hormone, made in the laboratory. It is the perfect substitute! Taken every morning before breakfast, replaces the amount of thyroxine our gland cannot produce adequately.

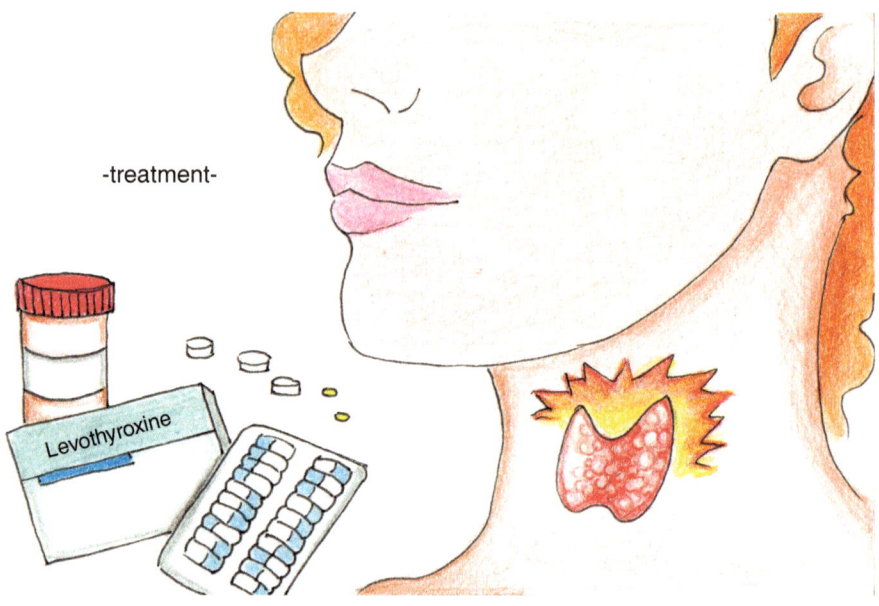

Levothyroxine is only available on prescription, and it comes in tablets or a liquid form; for instance, the new oral solutions of the medicine (e.g., Eltroxin Oral Solution or Levothyroxine Oral Solution) are even easier to take than tablets, since you can swallow it directly from the oral syringe or, if necessary, it can be administrated via a nasogastric feeding tube; you should take your dose on an empty stomach, usually before breakfast.

Of course, this treatment is not prescribed to everyone: it depends on how much TSH was found in the blood test results. The endocrinologist will evaluate case by case and prescribe the medicine in the right doses (which vary depending on age and clinical picture).

If necessary, pregnant women can also take it too!

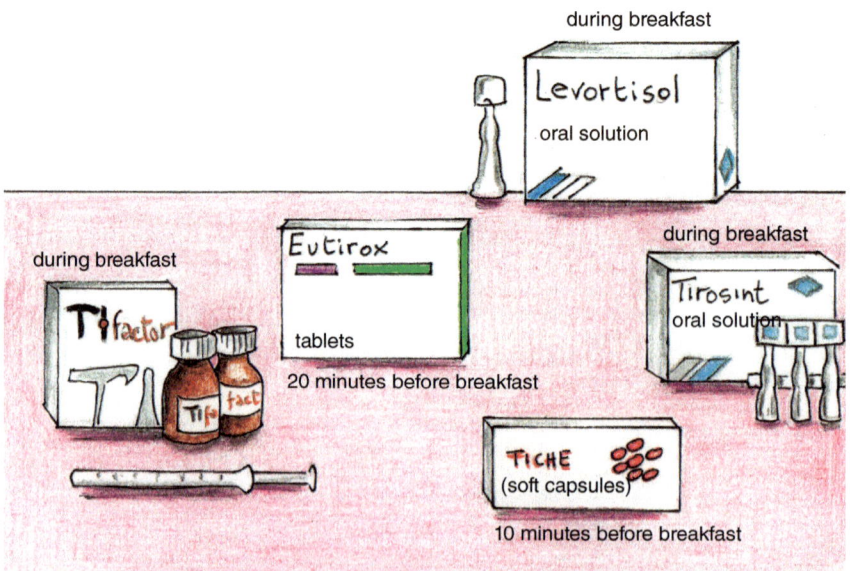

Actually, a woman who has hypothyroidism should take it before trying to have a baby! Those who take Levothyroxine must be careful with certain drugs and foods that could compromise the absorption of thyroid hormones hence the effectiveness of the treatment.

For example, you should eat soy in moderation and timing makes a difference; you should wait some time before taking other medicines as well as eating and that includes food and drinks that have soy in them Among the medicines that interfere with absorption, on the other hand, there are the so-called *proton-pump inhibitors*, or PPIs, which play a protective role when we suffer from gastritis, by reducing the amount of stomach acid made by glands in the lining of your stomach during digestion.

Overactive thyroid (hyperthyroidism) caused by Graves' disease, is treated with antithyroid drugs, such as **Carbimazole** and **Propylthiouracil**, which reduce the amount of hormones the thyroid produces.

Better results of this treatment are shown if associated with a Selenium supplement, which helps to decrease the presence of antibodies and improve the symptoms of inflammation that, can affect the orbits of our eyes, as we have seen in the second chapter.

In women who are expecting, this treatment can be interrupted, since generally hyperthyroidism improves in pregnancy and the symptoms lessen or disappear altogether.

## 5.2 Surgery

Medicines are not the only option in thyroid treatment. In some cases, we must undergo surgery to have our thyroid removed, either totally or partially.

Doctors may recommend surgery if you have a diagnosis of a *malignant* nodule or a nodule or goiter whose "nature"—whether good or bad—is hard to determine, despite going through ultrasound, fine needle aspiration, and scintigraphy.

You may also undergo surgery in case you have a goiter or a nodule that, although benign, is causing local symptoms such as compression of the trachea, difficulty swallowing and even speaking. Another case in which surgery is suggested is when hyperthyroidism does not respond to drugs.

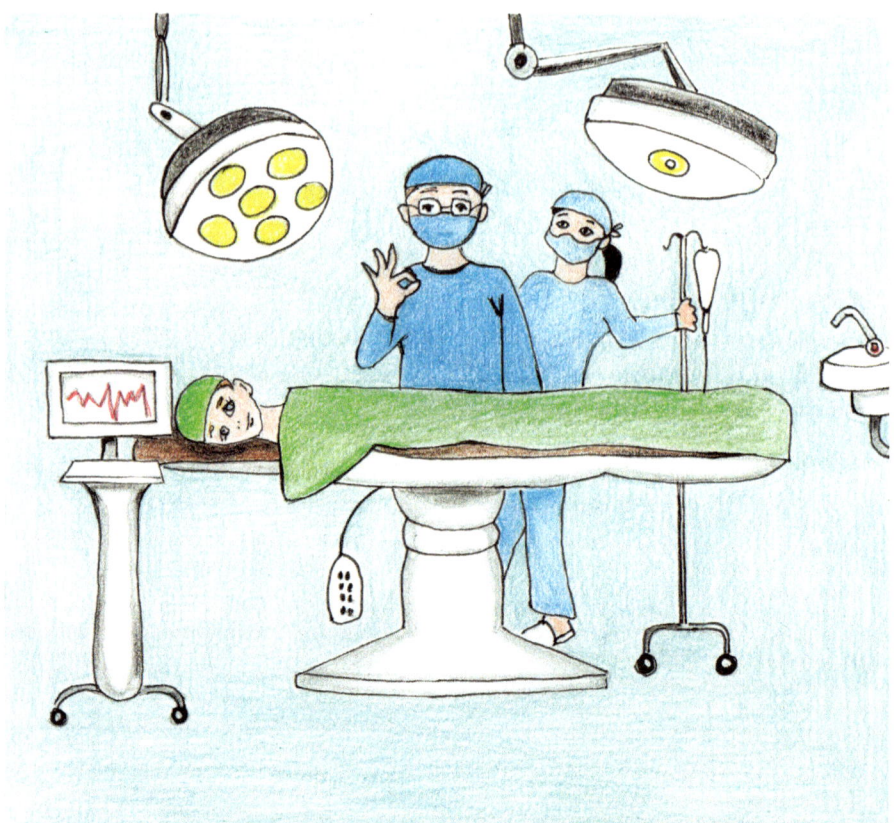

Based on our conditions, the doctor will assess whether it is appropriate to remove the entire gland (**total thyroidectomy**) or to leave a small part, leaving a little tissue on one side (**partial or near-total thyroidectomy**); removal of one lobe, so one half, of the thyroid (**thyroid lobo-isthmectomy** or **hemithyroidectomy**). Occasionally, he may extend the removal to the lymph nodes of the neck (**lymphadenectomy**).

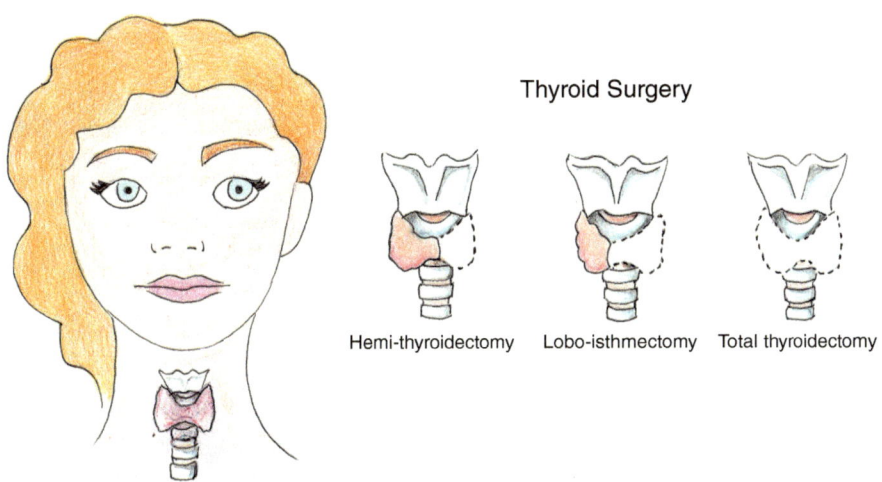

Thyroid Surgery

Hemi-thyroidectomy    Lobo-isthmectomy    Total thyroidectomy

The procedure may usually last from 45 mins to 3 hrs, using general anesthesia, so we will be completely asleep during the whole process; we will not feel any pain and we will not remember anything upon waking.

We can leave the hospital and go home, 1 or 2 days after the operation, some people may experience neck pain, and the small surgical scar will start fading or will almost become invisible when using sunscreen or local treatments.

If our entire thyroid is removed, its activity will have to be replaced by synthetic hormones, which is the same **Levothyroxine** that is used in case of hypothyroidism. Since our body can no longer make thyroid hormones, we'll need to take this medication every day to compensate for the lack of hormones usually produced by our gland.

When in good hands, the operation is a safe procedure and carries a low risk of complications: only in 5% of cases, episodes of bleeding can occur in the following hours, (which means that there's an excessive outflow of blood), and for this reason, we remain in the hospital one more day, where doctors and nurses will monitor our recovery.

In very few cases, there may be damage to the vocal cords, which causes a bit of hoarseness, if not even paralysis: a rather rare event. In other rare cases, the parathyroids, which are located right behind the thyroid, may be affected, and hypocalcemia may occur, meaning that calcium levels are lowered in the blood resulting in a tingling sensation in the mouth, hands, and feet.

This is a temporary circumstance too, curable with calcium and vitamin D intake.

It is also possible to resort to a **minimally invasive video-assisted surgery technique**, also called **MIVAT**, which is performed with the assistance of an endoscope via a small incision.

Unlike conventional surgery, MIVAT allows the minimization of any adverse event as well as the length of hospital stay, and aesthetic damage. In this case, the scar is really small.

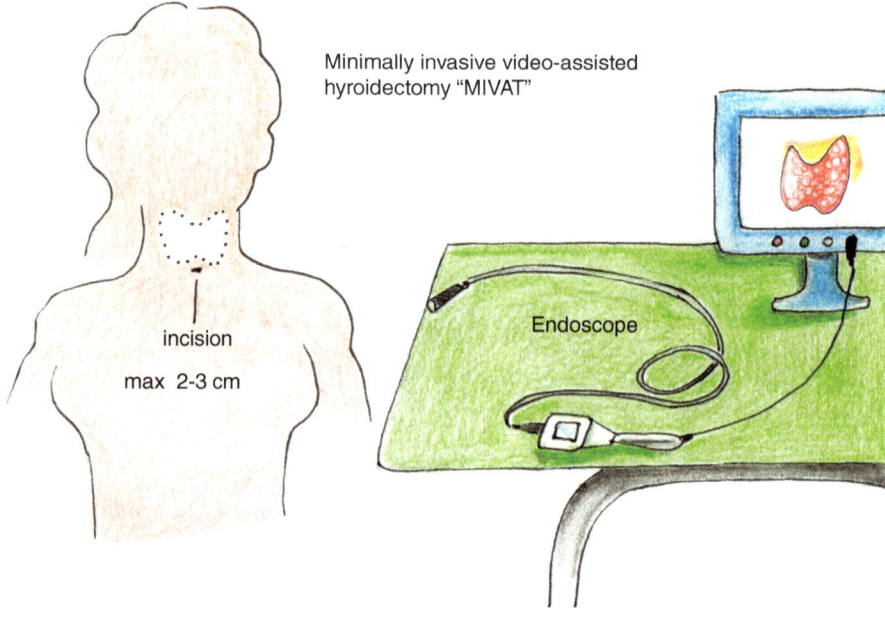

Minimally invasive video-assisted
hyroidectomy "MIVAT"

incision

max 2-3 cm

Endoscope

## 5.3    Radioactive Iodine Therapy Based on Iodine-131

Among the different thyroid treatments, we find Iodine 131, one of the two radioisotopes used for scintigraphy.

This treatment can be used mainly when antithyroid drugs are not enough to cure
hyperthyroidism or to ablate (destroy) any thyroid tissue not removed by surgery. In
the latter circumstance, the dose is much higher than the one used for scintigraphy.
It is not always necessary after surgery but it certainly helps in some cases. When it
is needed we are asked to stay isolated for a few days, as our body becomes
radioactive!

It is highly recommended to avoid contact with other people while waiting for our body to eliminate the substance.

For women, it's possible to plan a pregnancy; however, conception must wait for at least 6 months after the end of therapy.

If thyroidectomy was done due to a cancerous nodule, the only thing to do after treatment with radioactive iodine is to undergo periodic follow-ups.

It is good to know that the majority of thyroid tumors heal completely and that only a small percentage of these are hereditary. In any case, it's always better to carry out the right prevention with regular medical check-ups!

## 5.4    Mini-intervention Techniques Ultrasound-Guided

Thanks to technological progress, it is possible to achieve great results. Over recent years, percutaneous treatments (when a small needle is inserted into the skin) such as **ethanol Ablation** (EA) or **percutaneous ethanol injection** (PEI), **laser ablation**, and **radiofrequency ablation** (RFA) methods have been suggested to be effective outpatient procedures as they can obtain the same results as a surgical operation, obviously in a faster, minimally invasive way, with fewer complications and without leaving scars.

These treatments are ultrasound guided, and their goal is to reduce the volume of large benign nodules.

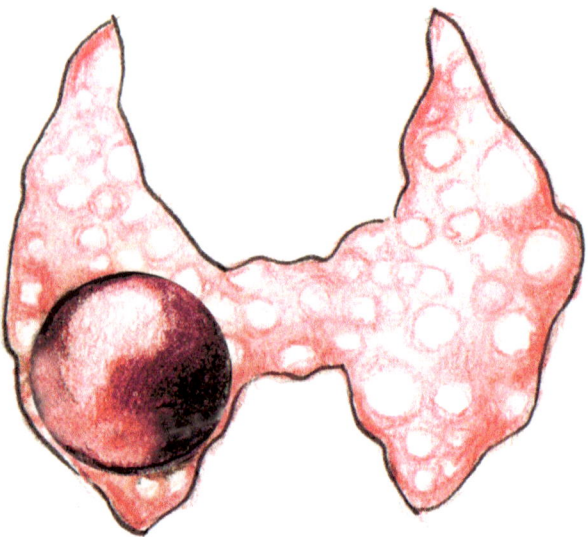

Specifically, in the **ethanol injection**, a needle is inserted through the epidermis into the nodule, under ultrasound guidance to locate the exact point where the puncture has to be done; the nodule is then emptied of the liquid portion and ethanol is slowly injected.

The procedure can cause a bit of discomfort and a burning sensation that remains for a few hours; rarely it can cause a bit of cough and hoarseness, but these symptoms tend to disappear in a few days.

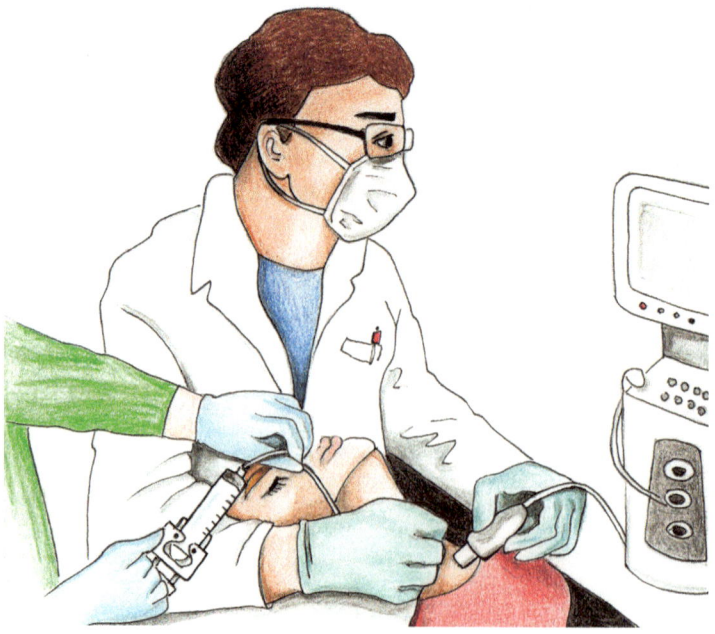

**Radiofrequency ablation** has been shown to considerably reduce nodule-related discomfort and cosmetic problems. The ultrasound-guided treatment is performed on an outpatient basis. An electrode that emits radiofrequency waves is inserted into the nodule and it is capable of "burning" the tissue. The constant ultrasound monitoring makes sure to avoid damage to the organs near our thyroid!

At the end of the operation, after the needle has been removed, a small plaster and ice packs are applied to the point where the electrode was inserted to prevent the formation of hematomas.

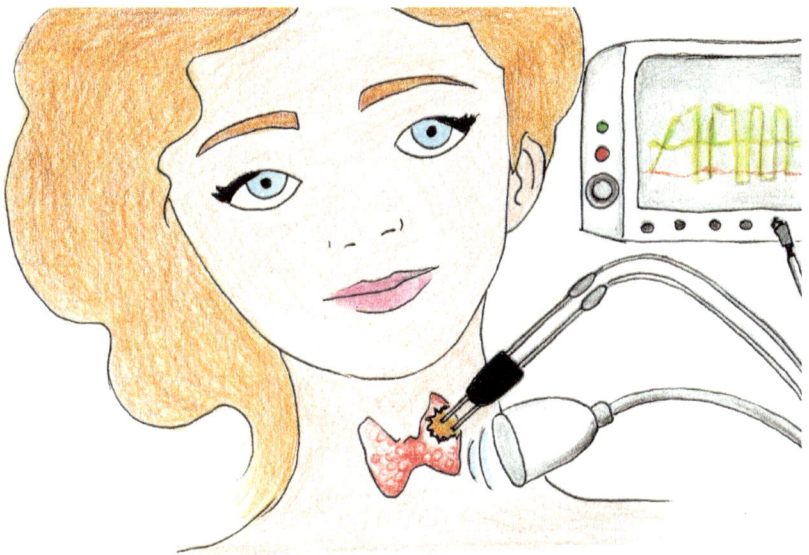

Radiofrequency ablation

In the **laser ablation**, the needle to be inserted is very thin leaving even fewer signs. An optical fiber is initially positioned through the needle and later withdrawn to leave the tip in direct contact with the nodule tissue that needs to be reduced, as always under the ultrasound guidance.

After the procedure, we can usually, go back home in 2 hrs or so without medications, although we will have to follow with periodic check-ups. The ablation treatment may be repeated after some time, just in the case of a double location in the other thyroid lobe which means that could be a valid technique for the reduction of large masses too.

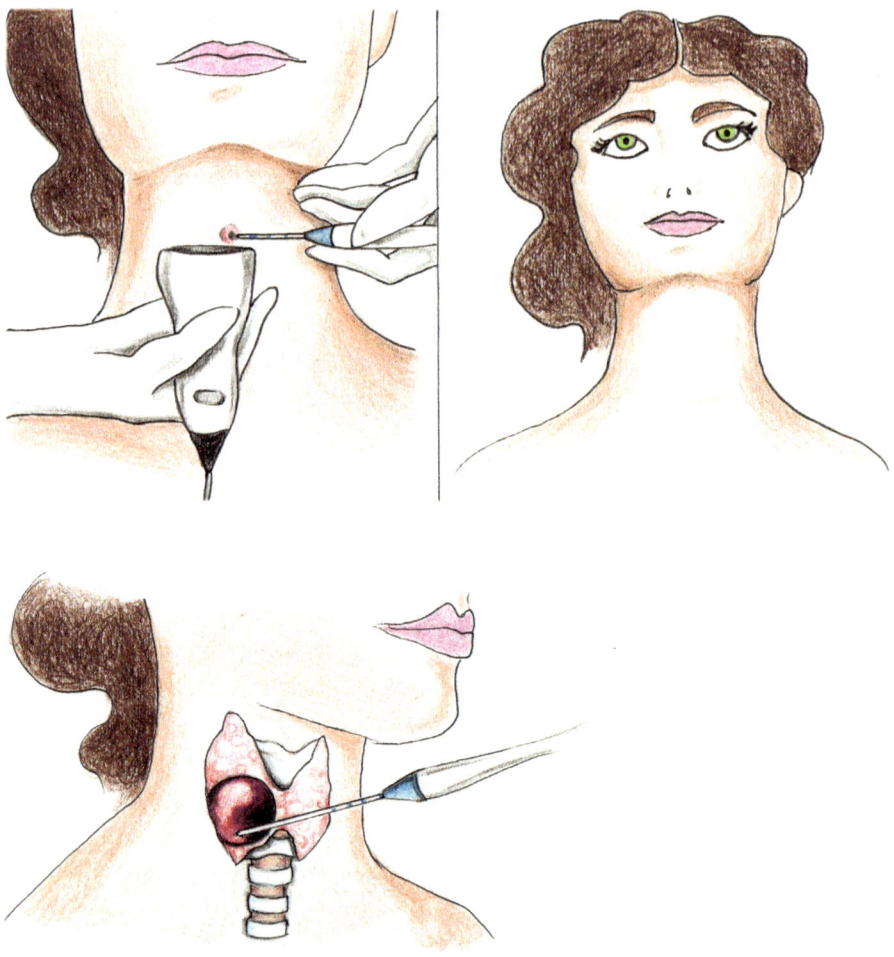

## 5.5   Surgery Done by Robots

Finally, there is another method to surgically intervene in a minimally invasive way, with limited post-operative pain and without leaving visible traces from an aesthetic point of view. This procedure was made possible thanks to scientific research and, of course, technological progress. The so-called **Robotic Thyroidectomy by**

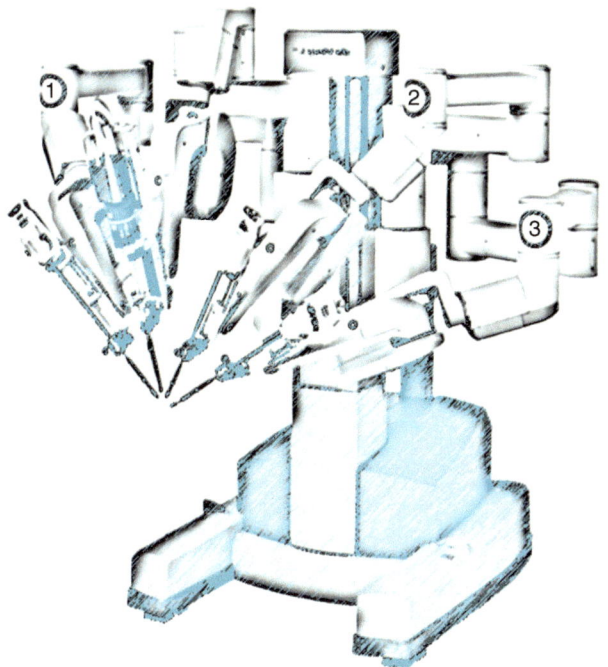

**Transaxillary Approach** (RATS) is an innovative technique that facilitates the total or partial removal of the thyroid offering improved cosmetic results on the neck, thanks to the **da Vinci robot** and its artificial intelligence. But let's understand more!

The transaxillary approach features a 2–3 in. incision concealed under the axilla, on the removal side of the thyroid lobe; if the entire thyroid must be removed, the incision will be done on the side where the gland is larger. A subcutaneous tunnel is created to allow the robot to reach our gland along with a tiny camera, which will provide a three-dimensional and magnified view. The robot, with its four robotic arms, three intended for the use of surgical instruments and one for the camera, will operate. Don't worry, da Vinci is controlled by the surgeon through a console equipped with a 3D monitor, two pedals, and two manual controls (like a video game!). With this minimally invasive technique, both benign and malignant lesions can be treated, although not all thyroid-related diseases can be helped by the da Vinci robot. The endocrinologist and endocrine surgeon will evaluate case by case. As already mentioned, among the advantages of this technique, we find reduced post-operative pain and improved cosmetic outcomes, since there's no scar on the neck and concealed in the armpit. The da Vinci Surgical System (DVSS) was designed in 2000 by *Intuitive Surgical* of Mountain View, in California, while the father of the Transaxillary Robotic-assisted Thyroidectomy surgery (2007) is the South Korean surgeon Woong Youn Chung.

✓ With specific diagnostic tests, the endocrinologist will be able to understand the health status of our thyroid and inform us if we suffer from any disorder, indicating all possible treatment paths where necessary. Some dysfunctions can be treated with medication. Levothyroxine, for example, is useful for those suffering from hypothyroidism, while methimazole and propylthiouracil are for those affected by hyperthyroidism.

✓ Levothyroxine is an exact replica of the T4 hormone and is recreated in the laboratory to intervene in cases where our gland produces insufficient hormones. The endocrinologist will assess each case and decide whether to prescribe it and at what dosage. Normally, it is taken by everyone with hypothyroidism, including pregnant women, as it does not cause disturbances. In fact, those under treatment who discover they are pregnant should immediately contact their endocrinologist because the dosage needs to be increased. As we have seen, during pregnancy, there is a greater need for thyroid hormone to support the changes in the woman's body and the proper development of the baby.

✓ Methimazole and Propylthiouracil, on the other hand, have the ability to block excessive hormone production, a situation that occurs when suffering from hyperthyroidism caused by Graves' disease or a goiter with hyperfunctioning nodules (hot nodules). Pregnant women taking either of these medications can safely discontinue the therapy because hyperthyroidism generally improves during pregnancy, and symptoms subside or disappear altogether.

✓ In some situations, such as when dealing with a malignant nodule or a greatly enlarged goiter, it is necessary to resort to surgical intervention for the total or partial removal of our thyroid. The surgery will be performed under anesthesia, and we will sleep throughout the procedure, waking up with no memory of it. It is relatively simple and has a low probability of complications. Of course, when the thyroid is removed, Levothyroxine must be taken because, with the gland no longer producing hormones, they need to be replaced artificially with synthetic hormone.

✓ If the entire thyroid needs to be removed, it is possible to opt for a minimally invasive video-assisted surgery technique called MIVAT, which is performed using a probe inserted through a small incision. This minimizes complications and duration, reducing the scar to very small dimensions. Among the treatment options for our gland is also Iodine 131, one of the two radioisotopes used for scintigraphy. It can be used as therapy in two cases: when there is hyperthyroidism that cannot be cured with antithyroid drugs or after thyroid removal due to the presence of a malignant nodule.

✓ A valid alternative to surgery is minimally invasive percutaneous techniques, represented by alcoholization and thermoablation; these methods are performed under ultrasound guidance to reduce the volume of large benign nodules. In alcoholization, a needle is inserted into the nodule with the help of ultrasound to locate the precise point for the puncture; the nodule is emptied of its liquid part, and alcohol is injected. In thermoablation (with radiofrequency or laser), an electrode needle is inserted, emitting radiofrequency or laser waves capable of "burning" the tissue to be reduced. In this case, the entire procedure is carried out under constant ultrasound monitoring.

✓ Finally, thanks to advances in scientific research and technological evolution, it is possible to surgically intervene using the Da Vinci robot and its artificial intelligence. This is the transaxillary robotic thyroidectomy, used for total or partial removal of the thyroid without leaving scars on the neck and with limited postoperative pain. The surgery involves making an incision of about 5–8 cm in the armpit, through which a subcutaneous tunnel is created to allow the robot to reach our gland along with a tiny camera, providing a three-dimensional and magnified view. The robot with its four robotic arms will perform the operation!

## 5.6 Conclusions

In these pages, you have encountered difficult words, challenging concepts that you probably read a couple of times to fully understand, procedures, and surgical interventions that would unsettle and intimidate even the bravest people. But don't be afraid! If you need to treat your thyroid, you will find allies in doctors and healthcare providers who are prepared and love their job, who will illustrate the right path to follow and guide you, day by day, toward healing. You will never be alone as you will start a journey with them; here, the only words you will find will be *attention*, *support*, and *care*!

However, beyond this eventuality, which I hope will never happen, remember that **it is important to know our thyroid**, to know everything about it, its functioning, and the diseases that could affect it, as well as recognize the symptoms of a possible hormonal dysfunction, etc.; it is also crucial to communicate them to healthcare providers and start a diagnostic-therapeutic process with a specialist.

Noticing a thyroid dysfunction in time is essential for the healing process because it allows us to start immediately any necessary treatment, preventing the disease from worsening and causing consequences to other organs of our body.

Never forget that **prevention** is fundamental and is a beautiful act of love for ourselves and our family.

So, let's take care of our gland, always adopt a healthy lifestyle and behaviors which are respectful of our body, as well as undergo medical examinations without fear or hesitation.

Let our butterfly fly free, away from dangers, obstacles, and diseases: only in this way, can we live together in health and wellness.